Citizenships, Contingency and the Countryside

T0347229

Citizenship became a buzzword in British politics in the 1990s, and under the Blair administration established itself as part of governmental and wider political rhetorics. The use of the term 'citizenship', however, ignores the lack of formal engagement of the public in politics and obscures how narrow state definitions of 'good' or 'appropriate' citizenship are. There is increasing interest in what citizenship means to the individual, and what constitutes a diffuse citizenship.

Gavin Parker argues that citizenship should be viewed more expansively and that an understanding of the role of culture and global change is integral to this aim. A citizen's actions, such as a consumer protest, should therefore be seen as expressions of alternative or postmodern citizenship. This splintering of action and conceptualisation should involve government and other institutions in rethinking how they recognise political action, prepare policy and themselves engage with citizens.

Citizenships, Contingency and the Countryside defines citizenship in relation to the rural environment. The book explores a widened conceptualisation of citizenship and sets out a range of examples where citizenship, at different levels, has been expressed in and over the rural environment. Part of the analysis includes a review of the political construction and use of citizenship rhetoric over the past twenty years, alongside an historical and theoretical discussion of citizenship and rights in the British countryside. The text concludes with a call to recognise and incorporate the multiple voices and interests in decision-making – voices which all affect the British countryside – and look at how participation, governance and land management need to be reconceptualised.

Gavin Parker is a chartered planner, specialising in countryside and environmental planning and management. He is based in the Department of Land Management and Development, at the University of Reading.

Routledge Studies in Human Geography

This series provides a forum for innovative, vibrant, and critical debate within Human Geography. Titles will reflect the wealth of research which is taking place in this diverse and ever-expanding field.

Contributions will be drawn from the main sub-disciplines and from innovative areas of work which have no particular sub-disciplinary allegiances.

Citizenships, Contingency and the Countryside

Rights, culture, land and the environment

Gavin Parker

Routledge
Taylor & Francis Group

LONDON AND NEW YORK

First published 2002 by Routledge

2 Park Square, Milton Park, Abingdon, Oxon OX14 4RN
711 Third Avenue, New York, NY 10017, USA

Routledge is an imprint of the Taylor & Francis Group, an informa business

First issued in paperback 2016

Typeset in Galliard by
Florence Production Ltd, Stoodleigh, Devon

British Library Cataloguing in Publication Data
A catalogue record for this book is available from the British Library

Library of Congress Cataloging in Publication Data
  Parker, Gavin, 1969–
    Citizenships, contingency and the countryside : rights, culture,
  land and the environment / Gavin Parker.
      p. cm.
  Includes bibliographical references and index.
    1. Citizenship—Great Britain. 2. Land tenure—Political
  aspects—Great Britain. 3. Country life—Political aspects—Great
  Britain. 4. Social change—Great Britain—Citizen participation.
  I. Title.
  JN906.P364 2001
  323.6′0941—dc21   2001031654

ISBN 978-0-415-19160-9 (hbk)
ISBN 978-1-138-97077-9 (pbk)

To the memory of F.G. Stokes

Happy are they who live in the dream of their own existence, and see all things in the light of their own minds; who walk by faith and hope; to whom the guiding star of their youth still shines from afar, and into whom the spirit of the world has not entered! The world has no hold on them. They are in it, not of it.

William Hazlitt, *Mind and Motive*

Contents

x *Contents*

Plates

Preface

In this book, citizenship as a concept is examined and applied in relation to rural politics, land and aspects of culture. In some senses this is a wide focus, in others it is a rather specific one. Citizenship is used as a cornerstone to explore changing society and culture and the changing countryside, and consequently also to explore the changing nature of citizenship itself. This is important, as citizenship can be seen as the way in which individuals engage politically and define the relations between individuals and all structures, not only nation-states and governments. As a consequence of this approach, the specifics of particular land uses, about economic activity, individual groups, or about any one particular locality in rural Britain, are intentionally not covered comprehensively. Instead, a new means of looking at citizenship – through the contestation of rural space and the brokerage of power between citizens and various groups of stakeholders – is introduced in theoretical and historical terms and through the use of case-study examples. The book should be seen as an excursus that holds important ramifications for 'rural' studies – even as far as problematising further the notion of a discrete category of rural.

It is argued that the need to look outwards and beyond any particular academic discipline is perhaps more necessary now than at any time in the past, because of the complexity that the information age brings (see Castells, 1997) and associated globalisation, but also from such notions as joined-up-ness that have been promulgated by academics and politicians during the 1990s. Such an epistemological and political context allows for a fresh commentary to be provided here about the category 'rural' and contestations over its form, content, meaning and trajectory(ies).

Following the quotation from Hazlitt used as an epigraph, the book's title has been chosen carefully; the book is about citizenship not only *in* the countryside but *of* the countryside. It is about competing definitions, imaginations and versions of history, alternative claims and attempts to hegemonise rural affairs and territorialise spaces. As a result of the approach taken, this book can be seen as a multidisciplinary synthesis, itself a reflexive hybrid drawing from political theory, cultural geography, social history, legal studies, town planning and land economy. These eclectic sources have common threads that are woven into this text: in particular, land rights and attempts to contest legal and customary land use.

Chapters of the book are in part outcomes of a number of research projects that have had an explicit focus on citizenship, that have assessed rural policies of one type or another, or that hold clear implications for the way in which we view citizen rights and responsibilities. They are also an outcome of personal research into citizenship, the historical development of rights, land use planning, and property and land issues. While I imply no claim about the conclusiveness or comprehensiveness of this work, I do begin to tie together some important themes. In so doing I encourage a more integrated and critical approach towards land-use studies. This should usefully involve explicit understandings and associations with culture, practice and cultural change in the 'freedom' and responsibilities that might be afforded to and taken by 'citizens' (and concomitantly demanded of other actants; see Hetherington and Law, 2000). In this sense, the stance acknowledges and integrates the contingent nature of both citizenship and land use.

The book is not simply an attempt to make sense of any particular political project that has been pursued in the UK and that has impacted directly on the rural. However, it is important that the past twenty years or so of UK politics has been witness to a resurgence of the notion of citizenship as a key ingredient of political rhetoric – a shift mirrored in the politics of other European and North American states. In the UK, Margaret Thatcher, John Major and Tony Blair have all made much of citizenship and of 'rights' and responsibilities during their periods of office. Blair in particular, with his 'Third Way' approach, including the notion of 'stakeholder capitalism' and the idea of 'engaged' citizenship, has been most explicit about how citizenship lies at the heart of UK government thinking. This may be reflected upon in terms of both how ministers think about their own actions (perhaps, as Klug (1997) suggests, to 'impact assess' in terms of rights; see also Barrow, 1997) and also how they wish the population at large to think about their role in New Labour's 'decent society' under a supposedly new social contract and an explicit return to community – even if a shallow or façadist one.

Added to this aspect of domestic politicking is the impact of globalisation in terms of the diminishing authority of the nation-state and its occlusion by supra-national institutions and challenges horizontally from the local and vertically by figurations or sociations (Elias, 1982; Hetherington, 1996). This situation, perhaps revealed to Blair by Anthony Giddens (see, for example, Giddens, 1998), may partly explain the 'control freakism' exhibited by New Labour in punishing any dissent within its own ranks and its 'strong state' attitude towards law and order policy. It seems that the 'engaged' citizenship envisioned by Blair, like the 'active citizenship' by John Major before him, is primarily conceived within tight legal, moral and cultural parameters. Perhaps such actions and constructions are consequences of a feeling of powerlessness, or even despair, at the rise of DIY or alternative politics and its often compelling consequences (McKay, 1998). Such social movements and the disillusionment with formal politics are not confined to the UK; even in Eastern societies, protests and DIYism are developing (see, for example, McCargo, 2000).

National politicians are beginning to reflect upon the global age, and it is plausible that they realise the possible implications for traditional modes of governance (Albrow, 1996; Urry, 2000). Such socio-economic changes, implied by postmodernity and post-nationalism, suggest that citizenship is a diverse and contested concept, with multiple formulations constitutive of many activities. It may be read in terms of process, of 'becoming', or striving to realise commonly beneficial goals as part of a 'politics of recognition' (see Gorman, 2000; Taylor, 1995). This represents an ethical, moral union as well as a cultural, economic and legal one (Van Gunsteren, 1998). It is this approach that is explored and applied to the rural in the following pages.

Acknowledgements

Over the course of writing this book I have worked at three separate institutions and in three very different departments, at Cheltenham, Surrey and Reading. Each in its own way has helped in the production of this text and I owe a debt to them for reasons of funding, forbearance and encouragement. Thanks to colleagues who have made comments and read parts of the text: Neil Ravenscroft, Mike Winter, Keith Halfacree, Julie Gore, Catherine Brace. I also take this opportunity to thank others who have encouraged and supported me in recent years, including David Crouch, Nigel Curry, Andy Pratt, Malcolm Moseley, Paul Selman, John Gyford, Judith Whateley, Amanda Wragg and Meiko Murayama. Lastly, thanks to my family and friends (not mutually exclusive from those already thanked!) for their love and encouragement.

1 Society, culture and rural land

Citizenship and the countryside

One of the frustrating things about writing about policy is that it changes. Thus change dominates the book, carrying as it does the theme of conditionality and contingency in terms of policy, but also in terms of culture and thence in terms of citizenship. In this chapter are set out the context and rationale adopted for approaching rural politics and land, particularly in providing an initial overview of citizenship as well as situating the countryside within wider social and cultural processes of change. It should be said at the outset that citizenship is conceived more expansively here than in many other texts bearing that label, although recently a widened and culturated application of the concept has begun to take hold (see Stevenson, 2001; Urry, 2000; Isin and Wood, 1999). This introductory chapter also invokes some of the key concepts that led to the use of a broader view of citizenship and that have been hailed as features of late-modern or postmodern societies. Therefore writing about citizenship and policy in the manner outlined carries an unavoidable double jeopardy. The book is first and foremost a discussion of citizenship, albeit a discussion of citizenship in a novel way and within a specific context.

Citizenship is closely tied to human rights and here that link is acknowledged, while the universality and fixity of 'rights' is challenged. Sedley (1997: 1), for example, accepts that rights are human inventions and are 'historically and ideologically the property of the liberal democracies of the West. . . . They are in essence the Enlightenment's values of possessive individualism, derived from the historic paradigm, which has shaped our world.' The modern and Western view of rights is one that is being contested, but for the majority they are being used and appropriated – traded for political and economic advantage by a multiplicity of groups using rights-claims, with all sorts of labels attached, to pursue and consolidate their interest in the face of (and drawing on) power. It is argued that culture in the widest sense develops citizenships while structures of various sorts attempt to control the ebbs and flows of cultural practice. Citizenship is a result of and a part of governance.

Citizenship in late-modern times should be viewed as multiple, contingent and subject to political manipulation by a growing number of agents at different scales. Rights of citizenship are not absolute, nor are they necessarily moral; however, they represent attempts to develop new exclusivities and can engender new conflicts as people struggle to reinvent both their world and themselves.

It is considered that aspects of citizenship may be examined through numerous policies, practices and texts beyond those given explicit debenture by the state and supra-national institutions. The label of citizenship is widely used, if not fully understood or investigated across disciplines. This provides one reason to explore its meaning and potential application in rural studies and planning. Citizenship is also an accessible vehicle with which to explore aspects of politics, culture, land and wider social theory. It is a useful concept in linking rural and urban affairs and similarly helpful in connecting different scales when looking at a particular policy area, locality or issue. (This links to the notion of 'action contexts' discussed by authors such as Habermas (1987, 1988, 1994).)

This project is undertaken in order to attempt to link a range of practices and changes in the rural that are involved in affecting the countryside on different scales and from disparate bases of concern. Making that link involves illustrating how governments use citizenship as rhetorical device, how agents use 'citizenship' and 'rights' as strategic devices and also how land, territory and space more generally are bound up with notions of citizenship and citizen claims of all sorts. In that sense they become action spaces for different interests to compete for 'citizenships' (see Goffman, 1967; Urry, 2000). A recognition and exploitation of 'citizenship' as a resource for manipulation is increasingly important in strategies of governance, control, resistance and public participation. It is considered that agents *qua* citizens on different levels of consciousness are acting or performing citizenship.

Importantly, then, the use of rights-claims and the impact of practice are underlined as important aspects of the performance of late-modern politics. Disparate and often dissonant claims for rights are important aspects of the brokerage of citizenship and often provide the impetus for conflict generally and land use and planning conflict more specifically. One of the repercussions of a risk-aware and rights-conscious society is that conflicts multiply. They are also increasingly mediated (see Routledge, 1997; Chesters, 2000), and the national and local states are finding it increasingly difficult to resolve emergent conflicts of interest. Instead, attempts to find consensus or build-in public involvement in policy and other political process is commonplace and often used as a means of obscuring, rather than altering, flows and networks of power (Forester, 1999). Citizenship tends to be portrayed narrowly in an attempt to restrict legitimate engagement as much as it purports to allow participation.

The countryside in the 1990s became the site and the category for increasing conflict and calls for radical change in regulation. The countryside has also maintained a fierce defence of amenity, and both traditional and new powerful interests have attempted to retain relative freedom from certain aspects of state control or the influence of wider social and cultural change (see Cherry and Rogers, 1996; Macnaghten and Urry 1998). Given that context, the book draws particularly on examples of conflict and protest in order to highlight the main themes of the book. That is to say, the book discusses land and 'post-citizenship', the claiming of rights, conceptions and uses of heritage and consumption in the rural, and the effects of imagined rurality on citizenship and practice. These themes are contextualised *inter alia* in debates about countryside access, hunting, land rights, planning issues and local heritage.

As stated, the use of space and the policy and politics relating to land underpin this work about citizenship. However, the book's scope includes an appreciation of citizenship (after Baudrillard, 1981) as implicated in the politics of symbolic exchange and a recognition that such 'citizenship' is part of the system of producing signs and commodities. It is also concerned with culture and cultural change, particularly as these relate to the use and reverence for history and heritage. An appreciation of heritage is becoming increasingly apparent in the contemporary UK countryside, particularly as rural space is increasingly geared towards consumers and leisure uses. I therefore investigate the role of consumers as contributive agents in rural politics (Urry, 2000). The connection between land and citizenship involves the interaction between place and the individual, and space and power relations through the crystallisations, or stabilisations, of group relations in the UK rural context. This chapter provides an introduction to the way that the main concepts are conceived and deployed and how the arguments concerning and analysis of citizenship relate to the contemporary countryside, while also outlining the overall content and trajectory of the book.

Culture, citizenship and rural policy

Citizenship has been discussed extensively over the past decade, especially within political science, legal studies and sociology. There has, however, been a conspicuous lack of attention paid to this concept in rural studies, even though the word has continued to be used in numerous books, papers and policy documents that are explicitly rural, or related to land use. Rarely is there commentary about what it might actually mean or an in-depth consideration given to the potential importance of theories of citizenship. It is also infrequently or only obliquely discussed how the concept might be usefully applied in many (rural) policy contexts (see Smith, 1989; Van Gunsteren, 1994, 1998; Macnaghten and Urry, 1998; Ravenscroft, 1998). In short, the term 'citizenship' has tended to be deployed without the actual or potential contents of the term being unpacked. It is used instead as shorthand for the useful or good behaviour that the nation-state, or particular government administrations, may require of individuals. This is a situation that on the one hand lags behind cultural and socio-economic change and on the other may victimise those who challenge extant 'norms' or legal definitions. This approach is reflected in government policy over the past twenty years, where citizenship has been a keyword of political rhetoric, but has lacked a deeper, wider or richer explication and application.

Citizenship is essentially reflective of distributions of power, although it is argued here that in order more fully to understand the flows and exercises of power, citizenship should be approached in an expansive, fluid way. Such a project extends into new areas and may offer novel insights; however, as is set out below, as many questions arise as are answered. This is particularly so when exploring what has been made of citizenship rhetoric by government and other interests. Definitions of citizenship constructed and maintained by the state, at more local levels, or beyond the state represent both contingent and conditional, yet powerful, definitions. It is suggested that the state attempts to constrain citizenship as much as to empower or activate people. The theoretical lens of this book is essentially a broader reconceptualisation of citizenship so that citizenship can be used to invigorate a debate about rural governance. This is achieved by expanding definitions of rights and responsibilities and by invoking culture, both global and local, as a key agent of political and economic change and resistance to change (cf. Cooper, 1998; Malatesta, 1974). It is poststructural, drawing on aspects of regulation theory and the sociology of translation as well as providing a political economy of the development of rights in the countryside (see, for example, Bourdieu, 1990; Giddens, 1991; Peet, 1998). Such an approach helps to analyse what the drivers and impacts of change in terms of individuals, groups, localities and nations might be. Therefore most of the book concerns itself in some way or another with how rights and responsibilities, as the core of citizenship, are being demanded and resisted in the countryside and how people engage with issues that are perceived to affect the countryside. In essence it is maintained that countryside politics, and the way that citizenship affects and is affected by politics, can be read as a war of manoeuvre – to paraphrase Mao Tse-tung, as a war without bloodshed.

Habermas (1994) makes the important point that citizenship is not necessarily a status that is tied to the nation-state, even though the nation-state has historically played the crucial role as the locus for the formulation and negotiation of rights and responsibilities of citizenship. Citizenship has been a creature of modernity, with the notion being tied to the national, to stability, progress and to ideas of commonality rather than of difference. There are exceptions in the literature of sociology and politics that have begun exploring wider ideas and applications of citizenship theory, notably the work of Clarke (1996) and Van Steenbergen (1994), and more recently that of Isin and Wood (1999), Urry (2000) and Stevenson (2001). Within those texts, key questions are posed; for example, how can a notion based on structure and exclusion be relevant in the global age? How does the condition of postmodernity impact on such a notion? Beginning to address such questions means that a more specific question needs to be raised: 'What is the relevance of citizenship theory to rural studies?' These issues are addressed using policies and examples rooted, in some sense, in the countryside, even though labels and demarcations of rural and countryside are in some sense provisional, arbitrary and partial in their implication (see, for example, Fine, 2001; Ray, 1999; Thrift, 1999; Murdoch and Pratt, 1993, 1994).

The notion of citizenship as homogeneous status or identity is challenged here. Held speculates, when assessing Giddens's analysis of citizenship, whether rights (as important components of citizenship) really are all they are made out to be. Are they in some way a sham whereby the powerful can control the proletariat? (see Held, 1989: 203; Giddens, 1984, 1985). Perhaps they are irrelevant in many circumstances; holding a right is not the same as exercising a right. It is argued that citizenship under modernity has indeed been an important part of state engineering of culture and society whereby rights and responsibilities become 'structural rules' (Featherstone, 1991; Archer, 1988; Giddens, 1979). The notion of cultural citizenship, extended from initial thoughts of Habermas (1994) and Turner (1994) and intertwined with normative constructions of legal citizenship, is deployed. Therefore the assessment of citizenship may be discerned to be more complex than the state simply enrolling the citizenry through rights allocations. Indeed, such a neat arrangement is rarely, if ever, the case.

It is argued that citizenship relates to the wider role of people and their activity in respect of the countryside. How rights and responsibilities are interpreted and developed in the context of the rural is an important part of understanding citizenship, conflict and the mediation of change. Rights and responsibilities (both legal and customary) come from culture and practice. Culture affects and practice is involved in the reflexive interpretation and therefore creation of rights and responsibilities, helping to shape activity and identity, resistance and compliance. Thus if resistance (towards political or other interests' attempts to foreclose debate, or impose narrow, interest-based claims) can be considered to be part of citizenship, then a very different 'frictional citizenship' can be conceived, one where dissent with plurality is honoured and so the 'politics of recognition' gains a more critical edge (see Taylor, 1995; Gorman, 2000; Smith, 2001).

When we think about the rural in this way it is not only those who live or work in rural spaces who should be considered, but also those who impact on the rural in some way. This view of the rural also applies when, for example, we consume the countryside or its produce. It implies a need to recognise the interconnectedness of people, issues and spaces. There is a moral dimension to such an approach in this regard. Sack (1993) makes reference to the complexity of impact that the actions of agents can have, indicating that how people behave and regulate themselves and others is, and perhaps should be, based on morality rather than imposed codes. Thus making governance in rural areas dependent more on diversity in those terms and not necessarily guided by a preoccupation with inclusivity in terms of place of residence or workplace may be an issue for reflection.

The notion of an integrated or national culture is itself a contested area of study. Such elements of the 'national culture' have been regarded as forming an important part of the 'dominant ideology' of the ruling classes rather than being reflective of true concerns and attributes of the people (see Malatesta, 1974; Abercrombie *et al.*, 1980; Mann, 1987). If anything, such concerns are heightened by increasing social and cultural diversity and reflexivity. In this respect the issue of nascency, becomingness and citizenship as process is discussed further in later chapters. Part of the process of change lies in the uneasy and dynamic relationships between group interests, localities and the central state; change into the twenty-first century involves the freeing of difference from enforced or muted integration of a dominant cultural and legal aesthetic born of an imagined rural.

The effect of actions through space and time and in terms of environmental, social, cultural and symbolic impact is traced as a form of policy analysis and political study. This is particularly pertinent, as explored later, in terms of empowerment debates and issues concerning different means of influencing or engaging in (micro)political action. Citizenship can be viewed as a cultural phenomenon in addition to a legal status or bundle of rights associated to the citizen. The ethnic or ethical community of citizens plays a strong part in determining their 'citizenship' in terms of day-to-day visible (and more opaque) community politics. In the latter chapters of this text, indicative, if partial, examples of community politics are included.

Rights are anticipated as being contracts between state and society that every person would value or need in order to maximise their life experience, or minimise the effects of mishap or misadventure – hence, for example, the arch-modern European Union and the Convention on Human Rights, which attempts to crystallise key rights much as a bill of rights would seek to do (and which has led to the UK introducing the Human Rights Act 1998) in order to provide certain guarantees that cannot be subverted by any particular political grouping (Charter88, 2000). In terms of planning, this focus on rights has caused consternation, as the ability to claim rights can be the dereliction of a responsibility on the part of the state – although it seems that such a bill of rights would have to be remarkably anodyne or general to avoid ingrained ideological bias (for example towards private ownership of land) and to prove acceptable to diverse communities of interest and political sociations. Even this line can be convincingly attacked as a liberal invention (see Sedley, 1997) and again as a reflection of extant (macro)power relations.

Rights, as mentioned, are conceptualised normatively as being those entitlements that are provided or guaranteed by the state and therefore can be claimed at law. Here, however, 'rights' are being considered more widely, where not only, or entirely, is it legal rights that fall within the scope of the citizenship 'envelope' as envisioned or designed by the state. 'Socially sanctioned activities' can be viewed as being a part of the local culture. Over the course of history, such activities have variously been eroded, developed or retained in different local circumstances and to differing degrees. Thus different citizenships may be sustained and the dialectic effects between habitus and social field envisaged by Bourdieu (1977, 1990) are mutated when difference is accepted. This implies that citizenship has been local as well as national, is processual and fluid, and possibly into the future will depend more on pleated and folded networks than on territoriality. In short, citizenship is viewed as contingent and multiple. Certainly Taylor (1995) sees the politics of universal equality and that of pluralism as potentially contradictory. A challenge (particularly perhaps in the changing country-side) is to find ways of mediating a social field that involves equal respect and respect for difference – a politics of recognition, as Gorman (2000) labels it in the 'urban' context. Such changes may be in respect of activities that take place (or which are restricted, such as hunting) or local provisions that exist or relate to characteristics particular to the person or group in question and are extended to particular people or wider groups (see Batie, 1984; Cohen, 1989). This is a high aim, as powerful interests would surely lose out if such a social environment were to take hold and implicitly were to lead to radical social change.

Imagining the rural

Policy plays an important part in forming and re-forming citizenship, as does day-to-day activity. Citizenship theory can be used as a conceptual tool for research, and already is enrolled to do service in justifying political policies. The actions of other powerful actors such as multi- or transnational compa-nies, or perhaps large landowners, can similarly be seen to shape culture, citizenship and rural policy. Throughout the book, the theme of history and historical examples are coupled with recent policy, acts of resistance and other overtly and sometimes unwitting political activity used to illuminate the theoretical aspects of the text.

The British countryside underwent significant change during the 1980s and 1990s culminating in numerous crises that are yet to be resolved – for example, over the fate of small farms, concerning food safety and produc-tion practices, and relating to the decline of rural services. There are fierce debates over appropriate rights of public access to the countryside and what leisure practices are appropriate in the countryside (e.g. hunting, offroading, mountain biking), and if, where and what forms of development and there-fore planning and other regulation should be formulated for rural areas. Recent constructions of the rural as being 'in crisis' are far from unique. Thirsk (1999) points out that many current rural issues facing society have been recurrent – particularly in terms of agricultural problems. Some may relate to fluctuating markets, while others are born of an increasingly reflexive, risk-oriented society or as a result of a trust in modern science that has not been entirely justified. It is clear, however, that change often comes about awkwardly, sometimes with hardship and almost always with resistance.

Serial crises and arguments over land and its use that have dogged the countryside are connected by larger or wider-scale pressures and changes in the global economy and society. They incorporate, *inter alia*, issues concerning regulatory frames and the effects of the Common Agricultural Policy (CAP), cheap agricultural produce from around the globe, new development at home, counter-urbanisation, and demands for a diverse leisure or a 'consumer's countryside' and for alternative lifestyles. They impact on issues of social exclusion, loss of basic services and other environmental considerations. Of course, some of these issues have been on the rural agenda for a considerable period. There are key changes in communication and constructions of the countryside – for example, the rise of media influence and the impact of the communications age more generally. This is allied with increasing reflexivity on the part of many sections of the public and the feeling of lack of control of 'internal affairs' owing to globalising forces. These together (and others omitted) seem to point towards the rural policy and politics field as being an increasingly complex and high-profile area – and one where, if claims of crises are to be taken seriously, national-level government may have seriously to reconsider the whole basis of economic, managerial and political regulation of the rural (see, for example, Rhodes, 1997). This involves the basis and flexibility of the (global) social contract and even what might constitute the 'stake' that a citizen might expect in a post-national and post-rural society, where distinctions between legitimate interests blur as does the rather artificial divide between town and country.

Instead of engaging with the specific dualism of urban/rural, this account concentrates on citizenship as comprising a global, local, multiple, fluid and contingent status and identity and involving practices of everyday life (Shotter, 1993; De Certeau, 1984; Crouch, 1997, 1999). This is a useful means to achieve a wider end: of examining how the state and administrations under the system of representative democracy attempt to cope with the actions of individuals and groups and how such communities and citizens themselves may 'perform the state' (Albrow, 1996) – and hence how structure and agency interact through the interplay of a 'postmodern' politics (see, for example, Routledge, 1997).

Notions of the rural and the countryside are often used as markers for particular spaces and by competing discourses. Such notions are considered to be of the rural rather than necessarily in the rural. This follows the idea that the rural is a state of mind more than an actual place or even than a set of particular practices (see Mormont, 1990; Woods, 1997b). Conversely, rural people and rural concerns, linked perhaps to particular land uses or amenity issues, are also part and parcel of wider societal dynamics. Rural studies, however defined (see Miller, 1996; Cloke, 1996; Winter, 1996; Crow, 1996), are increasingly seeking out the diversity that exists in the countryside and thus are challenging dominant constructions of the rural idyll (itself multiple and multifaceted). Such an appreciation of diversity and contingency in terms of identity and activity in terms of land and 'standing conditions' lends itself to the application of a widened citizenship theory regardless of categorisation from above (see Cloke and Little, 1997).

Rural imaginings are not necessarily bounded by historicity, place or law (see Blomley, 1994), especially in what has been termed the 'global age', where the impacts of technology and internationalised culture affect people and place in multiple ways (or in Robertson's (1995) terms are 'glocal', reflecting the way that localities perform or develop and relate to other localities and scales of governance on a global level). Rural places become legitimate space for competing interests and concerns to be played out and contested reflexively. Such a widening of scope is not new but when applied to citizenship it offers a new way of viewing practices taking place *about* the rural. There is an increasing tendency for new or alternative claims to be made about space, place and activity. For example, more and more groups come to view rural places as 'theirs', or, perhaps more powerfully, as no longer exclusive spaces.

Globalisation has had marked effects on society and is now impacting on the way that rural industry and social life are conducted. These impacts provide both localised responses and global impacts (see Giddens, 1998). It should be remembered that globalisation has been steadily developing. It was 1962 when Marshall McLuhan coined the phrase the 'global village' in reference to time–space compression (see McLuhan and Powers, 1992) and the phenomena of new communications, travel and intercultural mixing. It has been an ongoing and multifaceted process, with its roots in the Industrial Revolution and, of course, in the development of technologies of travel and communication. Bauman (1998) and Urry (2000) note that rapidity and mobility are key features of globalisation, linked as it is to the compression of time and space that enables change. The development of the railways and that of early communications technology are long recognised as being early features of time–space compression. In particular, the rise of the communications age during the 1980s and 1990s has been of unprecedented rapidity, bringing economies and cultures closer but with others left behind or abused by those able to exploit the global age.

An important facet of late modernity is said to be the rather haphazard or uneven connections between localities. Similarly haphazard or uneven are the relations between individuals and groups in the information age. As a consequence of the implied crises and pressures of 'glocal' change, some areas (see Robertson, 1995; Marsden, 1995) of the countryside are increasingly polarised. For some groups in the countryside the global or 'glocal' age can mean exclusion and perhaps the need to relocate, possibly into a communication/service node, most probably into the towns. Conversely, other groups have steadily traded places, giving the UK countryside a superficially steady if not increasing population level (DETR, 2000). Underneath, many people have left rural areas and other, predominantly middle-class, incomers have settled in the countryside. The networked and mobile middle class are countered by socially excluded groups who may have more in common with an American Mid-west farmer than their own neighbour (see Milbourne, 1996; Cloke and Little, 1997; Cloke *et al.*, 1998; Boyle and Halfacree, 1998). In this way important outcomes, or at least trajectories of change, involved in the ongoing process of globalisation (and particularly perhaps for the rural) have been both to de-traditionalise place and to de-territorialise space (Bauman, 1998; Albrow, 1996; Beck, 1998; Sack, 1986), while dominant images of the countryside scarcely change. These themes are readdressed in the final chapters of the book.

Alternative rights claims made in and over the countryside are given some consideration, particularly in terms of alternative land use and land rights/ responsibilities (see Halfacree, 1999; Parker and Wragg, 1999; Woods, 1997; Herman, 1993; Bromley, 1991, 1998). These can be viewed as examples of resistance and also of active citizenship, both of which relate to the three dimensions of citizenship embraced here: citizenship as status, identity and activity. An approach towards furthering such claims in the political arena is to appropriate legitimising texts or other discursively powerful tools. In this respect the Habermasian notion of a 'crisis of legitimation' (Habermas, 1988) comes into play whereby actors, as a result of developing understandings of contingency, uncertainty and risk, feel empowered increasingly to challenge societal norms and reflexively engage with capital, power and extant distributions of rights and responsibilities. Habermas and Luhmann (see Habermas, 1988: 130) view validity claims and 'social reality' as increasingly important – not because they are solidified but, to paraphrase Berman (1983), because when touched they 'melt to air'. In the postmodern or late-modern context, people, it is claimed, are not always guided by social or local norms but by their own experiences; the hyper-real life-world as discussed in Chapter 2.

Structure of the book: towards citizenships of the rural

The structure of the book including the chapter content is outlined in this section. As already introduced, citizenship and discourses of citizenship are viewed in terms of being important both in the countryside and of the countryside. Anderson (1983) noted that imaginations de-territorialise space and its communities of interest as well as actions or other physical manipulations of landscape (see also Sack, 1986). Therefore this text, while focusing on rural issues, argues that such concerns are arbitrarily located in that way. The issues of concern involve diverse spatial and social populations or communities of interest. Therefore citizenship and the rights of the citizen in terms of citizen action and participation and rural governance are implicit in the discussions and related to wider issues such as changes in class structure, conflict, counter-urbanisation and the new middle-class ('urban') influence on the countryside.

Much of the work done in the past relating to citizenship in political science and sociology is abstract and theoretical and has not been incorporated into rural studies. Little has been written that relates explicitly to space and place (exceptions include Smith, 1989; Fyfe, 1995; Urry, 2000) or which has been applied to public policy (although in certain areas this is less true – for example, in educational studies). The first part of the book consequently exhibits an emphasis on citizenship theory, although linkage is made to the rural context by examining historical and political events, policies and processes affecting citizenship with particular emphasis on rural space and land in those initial chapters. It is in the second part of the book, however, where specific and more recent countryside policies, events, actions and histories are applied and related to the theoretical and historical.

The text is at once exploratory, looking to expand and perhaps explode the notion of citizenship, though opting to retain the word at least, while simultaneously it seeks to provide examples of different scales, forms or expressions of citizenship in and around the rural. As mentioned at the outset, the book embraces a number of themes, including political projects of citizenship, land rights and reform, issues of public participation and of protest/resistance, consumerism and consumption practice. The role of the media and technology and the impacts of globalisation on rural society and in rural politics are also considered. These themes are cast against the frame of citizenship theory and viewed in the policy area of countryside policy and planning.

How places are regulated, how people are governed and how they conduct themselves are considered as important topics for citizenship studies and receive particular consideration in the second part of the book. The notion of citizenship is discussed in the light of the global and post-rural condition in Chapter 2. There are important issues that have remained unexplored in rural studies, such as the question of where rights come from. How are rights mediated and interpreted in rural localities? What is the role of the state in relation to the local and the global? How does culture inform and create 'rights'? And while I cannot claim that such issues are resolved here, or that they can be answered entirely through the use of citizenship theory, some indications are suggested. On a general level the theoretical sections pave out a political and social context of the 'post-rural', and more specifically show contrasting theoretical positions and approaches towards citizenship – that is, the communitarian and neo-liberal agendas illustrated in Chapter 3. This analysis is applied to examine rural policy and planning, and suggest how future directions might be enabled and understood.

I then examine the political and social context of the 'post-rural', setting the theoretical position and approach towards citizenship theory developed throughout the book. Chapter 2 deconstructs 'rights' and how people and structures in and of the rural are affected by global changes in economy(ies) and culture(s). Notions of 'what a right might be' are discussed and concepts drawn from cultural studies and social theory are employed to extend initial, normative conceptualisations of rights and wider citizenship. Particular emphasis is placed on the political project of successive governments in the UK (1979–99) in the third chapter. The way in which UK administrations have sought to develop and construct citizenship and policies invoking citizenship rhetoric are explored. This approach is then backed up by the case studies set out in Chapter 6. It is a useful opportunity to review the Thatcher/Major period (1980–97), to contrast and compare the ideas of that period with 'New Labour' policy and to transpose associated communitarian citizenship rhetoric into that discussion.

Attendant rights and responsibilities of citizenship are scrutinised in order to illustrate aspects of identity developed or stabilised through the state constructions of citizenship and through more organic citizenship action *qua* citizenship (see Kymlicka and Norman, 1994; Isin and Wood, 1999). The way in which policies affecting the countryside impact on its people and how the population at large reacts to actions taking place in the countryside are discussed. The use of citizenship rhetoric and underlying political philosophies are examined and related to land rights and rural policy more widely. Communitarianism and the New Left and liberal citizenship and the Thatcher legacy are discussed, along with some of the dualist and simplistic notions of active/passive and good/deviant citizenship that have been discernible in these political projects.

In general terms it is argued that in practice there has been little discernible difference in terms of the Major and Blair projects. Both administrations have viewed citizenship quite narrowly, while rhetorically sounding expansive. Even so, New Labour has begun to implement some interesting polices such as devolution for Scotland and Wales and creating an elected mayor for London, and possibly for other cities in the future. It is also exploring, or already legislating, on momentous issues: countryside recreational access (aka the 'Right to Roam'); the hunting issue, via the Burns Committee (2000); while the rural White Paper published in November 2000 contained numerous suggestions to attempt to redress declining services and the agricultural sector (DETR, 2000a) and to promote rural affairs to a higher position in government thinking. These policies are also given mention in Chapter 7.

The historical development of rights and citizenship in the rural is charted in Chapter 4, where an archaeology is begun with an historiography of citizenship construction, rights transfers, and examples of rural protest and claims for rights. The historical project of citizenship and state formation is discussed as part of an interspersed historical grounding used to illustrate how arguments over rights, and specifically in terms of land, have focused on remarkably similar themes over the centuries. These have, however, been prompted in very different social, economic and cultural contexts. This historical component primarily reviews the period since the English Civil War to the present, using citizenship theory to analyse change and the development of and resistance to the modern social contract. This analysis offers an historical review of the way in which rights and citizenship have developed in the nation-state with special reference to rural space, place and population. The assessment includes special reference to forms of historical rural protest and other key influences on rights in the countryside.

By these means, we become aware of how rights and responsibilities have been shaped and wrought in the countryside over time. Rights and citizenship offers a new angle of entry into a review of the well-charted changes and conflicts (over land) since the Agricultural Revolution (see, for example, Hill, 1996; Thompson, 1993). Historical rural protest and the influence on present rights accruing to particular groups in the countryside are discussed as well as key moments and challenges to the rural status quo such as the possibilities raised during and after the English Civil War and land reform and early planning movements in the latter part of the nineteenth century. Key issues relating to the private ownership of land and its exclusive use are woven into the analysis, and the issue of access to land for different consumption- and production-related purposes is discussed. The access issue and the related construction of trespass are introduced as a useful arena in which to analyse land rights and reflect on citizenship matters. The rise of present institutional influences over rural society and the rationalisation of land use and other activities on land are set out. This chapter also links present-day claims for land reform and governance using history and heritage discourses, as further discussed in Chapter 5, to their wider historical context.

Resistance is often cast as being a challenge to 'legitimate' power and authority, and while in many instances this may be true, it is argued here that many acts or forms of resistance should be reconceptualised as being necessary and positive aspects of governance and governmentality. Performing the state and performing corporations or other powerful groups is increasingly an important part of direct democracy (see Albrow, 1996; Monbiot, 2000), particularly when the state itself is being gradually occluded in world politics. This may open up new possibilities for rights and their interpretation and also for extended 'community' governance.

The second part of the book moves on to detail different ways of looking at citizenship, rights and contingency through particular policies or events that have been influencing the countryside and people in the countryside, especially during the 1990s. It does so by setting out six examples in three parallel chapters, each accompanied by further theoretical discussion. Chapter 5 opens with a discussion of issues of resistance to dominant rights while focusing on two case-study examples that link back to the historical and theoretical points made in the preceding chapters. The first details how a group makes use of history and heritage to energise and motivate actions relating to alternative land rights claims using a specific location and particular practices performed in Surrey, England, as a focus. The example also details subsequent engagement in cross-scale (local and national) politics and the effective use of 'heritage' in postmodern politics and as a key component of rural imaginations. Citizen attempts to participate in policy-making and rights brokerage are discussed in the later part of the chapter using research findings from a case study in the Wye Valley on the English and Welsh border. An example narrative drawn from research about the River Wye illustrates the way in which citizens, in the process of participating in policy and politics of the rural, can subvert dominant or centralised versions of citizenship and also how groups or individuals can be, and often are, marginalised in the process.

Chapter 6 makes use of a recent policy initiative (the Parish Paths Partnership scheme (P3) and other examples of policy that invoke rights and citizenship discourse) to tackle the very current issue of citizen empowerment and public participation in (countryside) planning (see Healey, 1997; Forester, 1999; Allmendinger, 2001). Constructs of the 'good' citizen and the 'deviant' citizen are challenged and notions of empowerment and self-help in the countryside are analysed. The P3 scheme is discussed as an example of 'active' citizenship and conflict in the countryside. It is argued that without more decision-making power, or at least input into formalised systems of governance, 'empowerment' strategies will remain largely façadist – that is, using discourses that pander to dominant local/national cultural imaginations of the rural. This argument highlights the differences between legal and national citizenship and local, perhaps customary, citizenships, and illustrates how state citizenship is enforced, often at the expense of attempts to foster local, organic processes of empowerment or capacity-building.

It is clear that many changes in the rural are not strictly of the rural, a point that could be made throughout agrarian and other rural histories. Numerous groups have historically blamed the 'urban' for the problems present in the countryside. Notably, claims that an 'urban jackboot' (Countryside Alliance, 1998) has been oppressing the rural are now in common currency among some 'traditional' rural interests – meaning that, supposedly, not only has the urban as place been adversely affecting rural areas, but that the mentality (the dispositions, imaginations and therefore attitudes) of urbanites has effectively marginalised 'rural' concerns. In Chapter 7 the example of the Countryside March is used to illustrate how hunting and other related, perhaps 'traditional', rural interests have exercised a form of active and 'postmodern' citizenship. Such an analysis is partial and reflects the cultural imagination of both the rural and the urban as particular stereotypical and distinct categories, imagined by some to be divorced from impact by, and impact upon, the wider economy and society.

The second part of Chapter 7 focuses on consumerism and the role of the consumer-citizen in global society and how consumers can impact on decision-making in the rural (see also Klein, 2000; Monbiot, 2000). The role of corporations, the consumer and that of the state in relation to citizenship are addressed. Sack (1993) underlines that consumer power can now affect space in important and lasting ways, and writers such as Shotter (1993) argue that the market can be a 'providential space' of freedom. The section reflects on wider impacts and intermediaries of change through consumer-citizenship and the media. The chapter also sets out the argument that the consumer is able to (and does) get involved in political activity. This is considered in terms of activities that either take place in rural areas, or are conceived as being 'rural' or 'environmental'. It is argued that such activity, for example politicised or ethical consumption, may also be considered to be facets of active or 'engaged' citizenship.

The example used in this regard is part of the actions taken over the proposed Newbury Bypass protests in the mid-1990s; the way in which protest groups used the market to claim rights and subvert the policy process is explored. The analysis extends to the environmental agenda and rights-claims of protesters; including the media role in influencing and supporting rights-claims or playing 'moral guardian' to demand performance from the state over such matters. These issues link in with the shifting usage of the countryside towards consumption space and a wider consideration of consumer society. Key concepts discussed are politicised consumption, the role of the 'consumer-citizen' and the commodification of the countryside.

In the concluding chapter, citizenship, land and governance in the post-rural context are assessed in the light of the prior sections. Widened conceptualisations of citizenship and the application of the text in terms of the rural and rural land are reflected upon, in particular in terms of globalisation and of consumption and consumerism. It appears that features of globalisation face strong opposition and resistance (examples being the Seattle protests in 1999, the Mayday demonstrations and the 26 September protests in Prague – both during 2000). Moreover, it seems that other protest, such as recent oil blockades in Europe, may be seen as a condition of the tension building up over outcomes of global economic flows. Although certain aspects of globalisation may represent more welcome or democratising advances (or could at least be extended and democratised, e.g. technological and communications advances), it is clear that on the economic and cultural levels there are already crises developing which are reflected in terms of conflict and economic failure. One such threatened backlash is that of nationalism with other associated, indicative forms of unrest such as growing outbursts of racist intolerance or arguments over national immigration policies – arguments that are currently climbing political agendas. Other potentially useful, but certainly destabilising, outcomes are political challenges made over a range of issues and claims by more and more group and individual interests, thus potentially engendering a more vibrant and transparent political culture.

The outlook for citizen empowerment and devolved powers of governance is reviewed, particularly in the light of past dominant political attitudes and the way that the state and particular administrations have constructed citizenship. The analysis and commentary on the development and current construction of citizenship in the UK are synthesised in the conclusion. This considers possible ways of rendering governance more flexible and relevant to (rural) people and how in a changing, restructuring (perhaps post-productivist) countryside, and a post-industrialist society, rights and responsibilities might usefully be reworked by Europe, the nation-state, local institutions and by individuals. It includes a review of where, how and why certain rights and entitlements are formulated, interpreted and tolerated at national and local levels.

The way in which citizenship and citizens (in and about the countryside) form or are reformulated and impact on the rural is central here, hence the notion of contingency, which unifies the key dimensions introduced in the book and which underlines the fluidity of the spaces and identities in and of the countryside. Reference is made to methods of making government (local and national) more accountable to local people and to ways of making the countryside a more openly plural and equitable patchwork of spaces and places, where a range of people and activities are tolerated. This, it is argued, may make for a more economically diverse, healthy and socially dynamic countryside. This should, following Dennison's conflictual but wise words in the wake of the 1942 Scott Report (MHLG, 1942; Cherry, 1975; Curry, 1993), deliver an actively used, appreciated, sustainable and cared-for countryside, for and by a wider range of (post-rural) citizens (see also Murdoch and Pratt, 1993, 1994). It is argued finally that citizenship in the global or post-national age is more a political tool or resource, open to widespread if uneven appropriation, than necessarily one of communal status or identity.

2 Unpacking citizenship

Introduction: considering citizenship

Citizenship is important *per se* because it is a key expression of the relationship between the state and the individual and of the individual to society – that is, it forms both a site and a conduit of the social contract. Official parameters of citizenship and associated rights are expressions of power rather than of rightness. Here the development and conceptualisation of citizenship in modern societies are dissected extensively and the components and mechanics of citizenship in liberal democracies are set out. This chapter builds on this exposition in assessing the way that citizenship has been envisaged, used and shaped by recent UK government in general terms, and particularly in respect of land, planning and the countryside. It sets out contextual citizenship theory and other related elements of social and cultural theory, plus aspects of socio-economic change relevant to citizenship and land.

Citizenship is important for rural studies because it provides a definition and a marker of status and rights/responsibilities. It is argued that citizenship is formed, performed and re-formed through activity; it is about participation in society and the role of the state and government. Therefore this resonates with debates over rural governance (Murdoch, 1997a; Edwards *et al.*, 1999) and the way that different interests exert control and influence in public policy. Citizenship may also be connected to everyday practice by embracing the notion of the 'post-rural' (Murdoch and Pratt, 1993, 1994) and reconceptualising citizenship theory in the postmodern context. These implications of and for citizenship are viewed along with aspects of objective and subjective identity. On one level, citizen rights and responsibilities shape who we are and what we can legitimately do (in the countryside), while other rights, responsibilities and practices are integral to rural politics and resistance.

Citizenship is deployed here as the central significatory label, if not meta-concept, largely because using it in this way enables the role and boundaries of government and interest groups to be examined with an explicit emphasis on the way that rights and responsibilities are altered, exchanged and negotiated through power relations. This approach does not preclude other analyses and concepts from being additionally mobilised; for example, network analysis and concepts of capital are usefully incorporated in the text (Bourdieu, 1984; Fine, 2001). As stated, taking a cultural approach towards citizenship, its boundaries and flows, means that the way in which power and influence are wielded can be included in such a commentary position. This approach is also used to critique political constructions of citizenship. Indeed, it seems appropriate that this stance be taken in the light of successive UK governments' use of citizenship, as detailed in Chapter 3, and notwithstanding the rise of other 'Third Way' administrations in the West.

It has been regarded as a hallmark of progressive modern societies that the more rights (civil, political and social) that a society affords people, the more 'advanced' or the more 'civilised' it is. Held (1989) implies that citizenship and claims on citizenship status cause friction. Thus rights deriving from citizenship imply change and contingency:

> Citizenship rights do serve to extend the range of human freedoms possible within industrial capitalist societies; they serve as levers of struggle, which are the very basis on which freedoms can be won and protected. But at the same time they continue to be the sparking-points of conflicts.
>
> (Held, 1989: 204)

Such a viewpoint is of course partial, culturally specific and based on a world-view that 'rights' are necessary preconditions for a civilised society. In addition to those types of criticisms are the substantive inward-looking contradictions about state power and control *qua* progression and civilisation and the boundedness of freedom (Sabine and Thorson, 1973; Rawls, 1979; Held, 1989). While freedom is an abstract concept, and at the opposite, after Foucault, absolute domination is not possible (Parker, 1999a; Parker and Ravenscroft, 1999; Foucault, 1977), the role of states should be to inform, engage and react flexibly towards cultural change and social need. That said, there are deeper aspects to be associated to citizenship, for example the way that citizenship is constructed and deployed discursively and expressed culturally and the way that individuals 'feel' citizenship (Clarke, 1996; Gorman, 2000). Such cultural citizenships are produced as part of the changing formulation of the cultural field (Bourdieu, 1990), which is increasingly influenced by global, environmental and historical factors.

The 1990s were witness to a marked increase in explicit attention to the rights and responsibilities of citizens. One might also argue that a countervailing inattention to the wider, deeper aspects of citizenship has also been present in many disciplines or by governments or commentators. This chapter seeks a re-examination; detailing how new conceptions of citizenship might be reflected in the case of the rural. The complex picture of global change is beyond the scope of this volume to explore fully, yet the impacts of such changes are important parts of the story of citizenship and contingency in the countryside. For that reason, a commentary on globalisation and the fragmentation of citizenship is interwoven through the chapter (and Chapter 3).

Recently, citizenship has come under intensified scrutiny, with some authors arguing for more research and for wider conceptualisations of citizenship to be contemplated (Falk, 2000; Isin and Wood, 1999; Van Gunsteren, 1998). It is overdue for citizenship, in the rural context, to be reconsidered in the light of radical social and political changes that have been taking place since Marshall was first writing about citizenship back in the 1940s and 1950s (Marshall and Bottomore, 1992). This necessary reconsideration is informed by the cultural turn in geography in the 1990s (see Peet, 1998) and can be associated with attempts to engage the wider public with planning and policy-making, particularly in the 1990s (Healey, 1997; Davies, 1999; LGMB, 1999; Forester, 1999). Why that is so may require some explanation: it became apparent that the label of citizenship was predominantly a shorthand convenience. There was work to be done in investigating the variances and problems that underlie such a broad and universalising idea as citizenship. There needed to be a further decentring of or destabilising commentary on current dominant definitions, or conceptualisations that restricted citizenship to a narrow set of (state-)defined rights and obligations forged through an explicit social contract.

In order to reconceptualise this seemingly most modernist of concepts, the invention of a more expansive flexible and contingent view of citizenship was necessary, to be added to the normative aspects of citizenship as being legally defined, or otherwise actively welcomed by governments. It is also possible to rethink citizenship in terms of *process* (Isin and Wood, 1999); and, further, to think about citizenship as a state of knowledge: of knowingness. If we creatively fuse these views of citizenship we can think of it, after Pred (1984), in terms of 'becoming'. In this way, the types of practice and activities that people engage in can be viewed more widely as being constitutive of citizenship.

Citizenship: status, identity and activity

The 'ideal' of citizenship has been conceived as one where all citizens are integrated into society and form part of that nation *qua* community. It has been the case that citizenship entry has effectively been open only to certain groups and that rights of citizenship have been exercised by those able to participate or claim such rights. This runs contrary to the demands of a modern democracy; but, as Giddens explains (1985: 202), such universality was not practicable in pre-modern societies. Citizenship was based on community or 'habitus':

In the feudal system rights were not universal, in other words, not applicable to every member of a national polity. Those in the various estates and corporations effectively belonged to separate communities, having different rights and duties in relation to one another.

It is clear, especially in terms of minority groups, that the 'ideal' has never been achieved. Arguably it is not achievable, and increasing awareness of difference and diversity makes it even less likely or desirable that the citizenship ideal as envisaged in the past should be pursued. It is worth exploring how a broader and more fluid construction of citizenship may be achieved and valued and a post-national citizenship be recognised.

Traditionally, the parameters of citizenship have been drawn narrowly and have, through sleight of definition, avoided, excluded or obscured many interests or alternative citizenship 'expressions'. There has never been an explicit acknowledgement by the state and political parties of the contingency and the radical alternative possibilities of citizenship. Similarly, the formulation and contestation of citizenship have not been adequately explored. Clarke (1996) suggests that there has been an avoidance of 'deep citizenship'. A divide exists that is observed, largely in mechanical fashion, between the boundaries of legal citizenship and new forms of claims. Isin and Wood (1999) partly make this link by thinking about citizenship as process, but in similar fashion to the Marshallian model appear also to conceptualise such a process as being one-way or evolutionary. Added to this, it is argued that consideration of forces using 'rights talk', or claiming citizenship in the juridical sense, are also placing a strain on modern nation-based citizenship. The expression of citizenship is mutating such that actions and enforcements are based more on mediated information and cultural determinants from above or from below the level of the national state as well as being promoted as a vehicle for group or network politics.

Here it is asserted that there are multidirectional forces that render citizenship both conditional and contingent: subject to continual contestation and renewal. The notion of citizenship as traditionally conceived has been a cornerstone, or unifying concept, for a progressive society to build upon. Heater (1990: 285) presciently noted that

> as more and more diverse interests identify particular elements for their doctrinal and practical needs, so the component parts of the citizenship idea are being made to do service for the whole. And under the strain of these centrifugal forces, citizenship as a total ideal may be threatened with disintegration.

It is likely to be put under considerable pressure, as Heater (ibid.) has noted:

> Citizenship as a useful political concept is in danger of being torn asunder; and any hope of a coherent civic education left in tatters as a

consequence. By a bitter twist of historical fate, the concept, which evolved to provide a sense of identity and community, is on the verge of becoming a source of communal dissension.

This view is somewhat apocalyptic. The central idea is, however, that as a more diverse and culturally fragmented society develops, the servicing of the needs and aspirations of those people becomes equally diverse and problematic for the state. It follows that the political construction of citizenship, by the state, should both expand and integrate, accepting a wider definition of the 'good' citizen (see Parker, 1997; DoE/MAFF, 1995; Ravenscroft, 1993). Total assimilation, as with domination, is not actually possible. This is reflected in a number of ways, but one important projection may be discerned by examining state attitudes to dissent or protest (see Sibley, 1995; Parker, 1999a). Political demarcation constructs citizenship conditionally and expediently; such processes are inherently linked to the deployment of different types of power. It is contended that rights distribution and, more widely, citizenship is a contingent crystallisation of power.

T. H. Marshall and the typology of rights

Marshall is now regarded as the academic who first formulated citizenship theory, and his work is still widely referenced fifty years after the publication of his seminal *Citizenship and Social Class* (Marshall and Bottomore, 1992; see Falk, 2000; Van Steenbergen, 1994). Marshallian citizenship, as developed in the 1940s and 1950s, involved the full membership of a community, entailing participation and comprising equal rights and duties, liberties and constraints, powers and responsibilities (Marshall and Bottomore, 1992). His view, based on the development of the welfare state and prior extensions of civil and political rights in the UK, was evolutionary in nature, implicitly viewing rights as milestones on a progressive journey, one that was primarily aimed at curbing the impacts of the free market. Marshall argues that social rights came about as a result of civil and political rights (such as stepped extensions of the franchise) having been accepted as legitimate by dominant legal and political authority (although not enshrined in a formal constitution and therefore liable to revocation; see Klug, 1997; Giddens, 1985).

Marshall's evolutionary conception of rights development has been criticised by more contemporary writers, because it is clear that many rights came about through a process of political lobbying. King (1987), however, seems to adopt the same Marshallian view of the evolutionary and rolling nature of rights gained under the welfare state, but it is questionable whether rights, once gained, become permanently entrenched and adopted as part of the nation's political culture. It is not at all clear that the rights gained were an organic or irreversible development within society. Held (1989) emphasises class struggle as being necessary to bring about the development of citizenship gains, while Giddens also argues that these rights were fought for via class conflict, and underlines the idea that rights require continual defence: 'rights once established can come under attack or be dissolved, and the history of other states across the face of the world demonstrates clearly enough that the categories of citizenship right form substantially independent arenas of struggle' (1985: 320). This view certainly suggests that citizenship may be both conditional and contingent. Giddens takes the view that 'real' rights or rights that 'make a difference' need to be defended rigorously. This supposes that the rights of a citizen are actually effective *de facto* towards equality of outcome. It can be argued that the emplacement of such rights forms a part of the social contract or 'gift' relationship (see Parker and Ravenscroft, 1999; Mauss, 1990) implicit with hegemonic (or territorial) trade-offs.

Additionally, some social rights have turned out to favour the more affluent in society even when they have been intended for those who are in need of socio-economic support, or for structures to be maintained to allow for fairer 'entry'. This can be read as the main reason why such rights remain in place. Examples include universal benefits systems/entitlements and their role in assisting already affluent families, the land-use planning system in maintaining the rights of those with property interests (Allison, 1975; Ambrose, 1986; Monbiot, 2000), the provision of amenity space in the countryside for middle-class and largely white use (see Curry, 1994; Kinsman, 1996), and the state education system in providing free education for more affluent families. Plant (1994: 186) neatly outlines how all rights, including property rights, infringe on freedom:

> Taking property rights as given in our society in which there are virtually no unowned resources restricts freedom of non-property owners to exercise their liberty. Hence the real question is not about the infringement of liberty. The question is rather whether, for example, the right to the means of life has priority over the unfettered right to property.

Smith (1989: 148) claims that 'citizenship theory provides a vision for the transformation of society which rests neither on the overthrow of the state nor on the sanctity of the market'. The progressive transformation or role that Smith envisages is based on a democratic development of the constitution of citizenship rights and responsibilities. In a postmodern context, however, typologies of citizenship rights when applied to the full range of UK rights show tensions between the categories of rights and between the groups who hold those rights. While the mantle of citizen is notionally shared, the same ability to exercise or enjoy rights is not equally shared. It is also the case that citizenship as status has never been shared even by all the inhabitants of a particular nation-state.

Citizenship can be viewed as an 'envelope' of rights, responsibilities, entitlements and obligations. Such an envelope is viewed here as enabling and constraining aspects of identity. Different political projects will demand differing constructions of citizenship; and a different envelope of rights and responsibilities, entitlements and obligations tends to be devised as part of that project – hence calls for a Bill of Rights in order to guarantee certain rights and obligations (see Charter88, 2000; Blackburn, 1997) and the recent implementation of the European Convention of Human Rights in English law through the Human Rights Act 1998 (see Blackman, 2001). Such projects may also influence national and local – near and distant culture. The notion of this 'package' or 'bundle' of rights lies, as discussed in Chapters 3 and 4, within the concept of the social contract (see Clarke, 1994).

It may be observed that there are differing facets to 'citizenship': Kymlicka and Norman point out two of these concepts (1994: 353):

> there are two concepts which are sometimes conflated . . . citizenship as legal status, that is, as full membership in a particular political community; and citizenship-as-desirable activity, where the extent and quality of one's citizenship is a function of one's participation in that community.

These relate to legal state-formulated citizenship and the wider citizenship derived from 'community' membership and legitimate participation in society – this is what has been defined and portrayed as 'good' citizenship. The legal-based definition of citizenship is one that began as a non-economic concept, with the elements of citizenship being unconditional (Dahrendorf, 1994). The latter element of citizenship as activity is, however, based not only on the legal, but importantly on the social and cultural. It can be argued that 'citizenship' is shifting more towards the latter, and the agency component of citizenship is being exhorted more and more, as seen in governmental rhetoric of 'active' and 'engaged' citizenship.

The third aspect incorporated here is citizenship as identity. This is complex: identity is multifaceted and multiform although it can be constitutive of attributes necessary for recognition by others; implicitly forming groups of affinity, or for constructing the 'other'. Identity is also unstable, being formed through a range of 'discourses, practices and positions . . . identities are about drawing on the resources of history, language and culture in the process of becoming rather than being' (Hall, 1996: 4). As such, identity and its counterpart, representation, can be an integral part of a process of 'becoming citizen', for example in terms of status, identity and participation/activity. Identity is something that, in contradistinction to status, is subjective. The identity aspect can also be formed by state attempts at citizenship construction, but may also be oppositional, or at least resistant, to the dominant construction.

It is argued here that rights represent part of state (and occasionally suprastate) attempts to regulate society and individuals as well as being features of investment in society by citizens themselves. This device is important.

Structure exists only through the activities of agents, both human and non-human, and the view of structure varies with domain, social field or world-view of the actor. As a consequence, citizenship construction can also be viewed as being part of attempts on the part of agents to regulate the state and other groups in society – in Albrow's (1996) terms, 'performing the state'. This includes the manipulation of space and objects in space through the maintenance of territoriality (Sack, 1986). The interrelation between space and citizenship is further explored later in the chapter.

Different political ideologies propound varying roles for the state and the individual, and implicitly require differing distributions of rights and responsibilities. The delineation of appropriate activity and projects of national identity construction are part of this – for example, where to walk in the countryside, or the cross-curricular programme of citizenship education taught in state schools (Herman, 1993; Cooper, 1993; Institute for Citizenship, 1998). The libertarian view of citizenship, at one extreme, does not accept anything more than a minimal state and a minimal citizenship enabling certain civil and political rights – dependent on their particular characteristics and outcomes. Set against this, the social democratic ideology has been based broadly on extending rights wherever there is 'need' – that is to say, where groups have successfully convinced receptive (or embattled) government of the legitimacy of a rights-claim based on moral, ethical, environmental or other 'social' grounds.

A proliferation of citizenship types has been proposed in recent literatures. Notably, environmental and social theory texts have *inter alia* identified ecological citizenship, global citizenship, cultural citizenship, post-citizenship, sexual citizenship, post-colonial citizenship and consumer-citizenship (see Van Steenbergen, 1994; Bulmer and Rees, 1996; Isin and Wood, 1999; Falk, 2000). Marshall provided a template that reflected his era. Isin and Wood (1999: viii) outline why citizenship requires further examination in the postmodern or late-modern context and how it may be of particular relevance to (rural) space:

> The focus of early citizenship was the specificity of particular rights and freedoms, which were to reside in the individual. The actual practice and process of those rights were only ever conceived in the abstract. Moreover they were not conceived with any recognition of the relevance of space . . . Marshall's work does not go far enough.

One of the suggested labels for new citizenships also segregates citizenship spatially through specifying 'urban citizenship'. Lefebvre (1996) implies that there is some distinction between citizenship in different spatio-cultural contexts. It is also the case that Marshall's work did not fully embrace issues of culture and the local. Lefebvre's work implies some essential difference between the urban and, as a corollary, what may, or may not, constitute 'rural citizenship'. Certainly this may be to do with aspects of 'group' rights or in terms of imagined, experienced or 'known' elements of citizenship and internalised identity *qua* citizenship (relating to the idea of habitus). Although citizenship continues to be defined by the state in terms of territorial affinity, space is being replaced in primacy by time. There is certainly a relationship between, for example, the experience of space and activity in relation to time and the experience of becoming through time (see Adam, 1995; Nowotny, 1994). Falk (2000) argues that time, in terms of looking towards the future for solutions, is the medium now for constructing a global, compassionate citizenship. This perhaps relates to issues of deliberation in terms of decision-making that are becoming more important in policy circles (see Bloomfield *et al.*, 1998). According to Falk, present structures are unable to deliver such a citizenship. Space and citizenship is picked up again as a continuing thread later in this chapter.

There has been little said in the literatures about where rights actually come from, and even how they are used. There also seems to be little written about rights and localities and how local rights and responsibilities fit into wider notions of citizenship. There is a rich vein of work here that has only just been opened up. Isin and Wood (1999) have called for this to be prioritised in research terms, and assessments of how citizenships are being re-formed and contested in top-down and bottom-up sets of processes are needed. Citizenship is being moulded by globalising forces and conversely by other, more local or alternative cultural reflexivities. Of particular interest here is how those relations are being moulded in rural contexts. Assessments of policy and other change in terms of impacts on rights and responsibilities (and even wider forms of evaluation) could become important decision-making tools in the future.

There are a number of issues and questions that arise from new conceptions of citizenship. Some of these are discussed in this text, while others are simply stated as topics that require further exploration. It is noted, therefore, that many important aspects in terms of identity (e.g. gender, ethnicity) are far beyond the scope of this volume, as are very many types of participation. Several complementary aspects are looked at in later chapters, but predominantly the participative 'citizenship as doing' aspect is analysed, especially in relation to space and with a focus on rural land.

A question mark stands over why a conceptual divide has persisted between certain forms of participation in a democracy and other types of transformative action – often constructed in a bipolar sense as 'good' participation and 'bad' protest. Such a construction is partly enabled by the deployment of polarising discourses that underpin the positions of particular groups or sociations in whose interest it is to delegitimise particular practices or forms of political action. Such mobilising discourses are diversifying as part of reflexive modernisation (Lash and Urry, 1994). As part of this 'postmodernisation' or loosening of citizenship, the consumer-citizen and the role of the cultural imagination in the construction of rights, rights-claims and action may also be incorporated. Before we further examine the constructedness of citizenship, legitimating discourse and the new challenges that confront 'nations' in the late-modern or postmodern context, it is necessary to unpack and appraise the components and development of citizenship and associated theory. This chapter is not, however, intended as a legal treatise on the topic, but rather an attempt to combine elements of legal theory with cultural studies.

Citizenship theory and legitimation

In consequence of processual deployments of power, various groups and individuals have gained, lost, regained and redefined different rights achieved through a process of political and economic brokerage and through class struggle. Indeed, this process can be viewed as largely having been a competition for the legitimate bearing of rights (see Giddens, 1985; Isin and Wood, 1999) and one reflection of the use of different forms of capital (see Bourdieu, 1990, 1991; DETR, 1999). In this respect, Chapter 4 provides a narrative on rights over land in the rural context.

In one sense, all 'rights' are ascribed or 'owned', in the same way that all land is 'owned' by someone. Kymlicka and Norman (1994) claim that the upsurge of interest in citizenship during the 1990s is a natural progression from the political philosophy debates over justice and community membership during the 1970s and 1980s respectively. Tönnies's work is important in terms of the distinction between community and association (*gemeinschaft* and *gesellschaft*) (Tönnies, 1963) and therefore with the formulation of citizenship. The importance is that, increasingly, associations or groups have battled for rights. This is the case even though group rights, in the legal sense, do not exist; *rights, instead, are universal but conditional.* A group must therefore convince other groups or legitimating authority that their rights are legitimate, necessary or at least justifiable 'trades' against other rights.

Rights may be conditional upon landownership, age, gender and cultural practice (a theme of this work is leisure and consumption practices in the countryside). Rights can be as much about social exclusion as they are about social inclusion. The types of groups in question struggle to assert a claim to have their existence, as well as their identity, recognised and the scope of their legitimate activities acknowledged in order to claim rights and to ensure the enforcement of existing group and state responsibilities towards them. Such claims are investigated in later chapters. It is clear, however, that citizenship is under continual political pressure; its shape(s) reflect(s) the contours of power and the deployment of social, cultural and economic capital.

Citizenship in the theoretical model developed here can be seen as a conceptual frame to help explain how people and practices are affected by flows of power and, for example, history and heritage, media spectacularisation, global flows of information, local introversion – and, importantly, how people interpret and reflect such cultural and social change in terms of action, imagination and rights-claims. In widening the idea of what a right is, and beginning to examine where 'rights' come from, some may argue that the concept of citizenship is misused. The addition of such dimensions, however, requires politicians and policy-makers to reflect on where and how rights and responsibilities develop. Both legal and cultural definitions of citizenship are included as part of the review of citizenship below.

What is a right? De jure, de facto *and 'nascent' rights*

According to Hohfeld in his *Fundamental Legal Conceptions* (1919), the legal conception of a right is 'a claim on an act or forbearance from another'. As will be discussed, it is how and by whom a claim can be procured, articulated, recognised and justified that is the crucial question affecting this discussion. The legal definitions also indicate how rights affect and are affected by activity or practice. Many rights, or claims to rights, are claims to liberty (and also claims to power). One of the obstacles to change, in terms of the structure and interpretation of (property) rights, is the evolution of accepted claims by legal systems; this is where rights, as definitions of the legitimate and the illegitimate, are stabilised. Other notions that may be labelled as 'nascent' rights are also included in the wider discussion here. These are the aspects of the cultural field and habitus that are born of change and are yet to be (if ever) presented as claims on behalf of an individual, group or class at the national level, but may be accepted locally, or across a community of interest. Hence they may or may not be constitutive of rights-claims.

The legitimacy of rights and rights-claims is key. It is the construction of claims that are deemed acceptable, owing to the dominance of a particular worldview or discourse, that constrains the capacity to express a claim-right (see Becker, 1977) or to gain sufficient exposure and support for a claim. There is an obvious cultural dimension to this, the cultural field may be subject to constant change and certainly differs from place to place just as, conversely, other aspects of culture remain recognisably stable or durable. Cooper (1998) goes so far as to claim that the social contract has been replaced by the cultural contract, where there is an imaginary settlement between members of a community to accept a particular set of governance relations and practices.

Such social or cultural contracts are rarely without resistance and opposition, and in a de-traditionalised context (see Giddens, 1998, 2000) may be exacerbated either to the detriment of the dissenter or advantageously in opening out new possibilities for introducing new or reclaimed tradition (Parker, 1999b). Both points make anomie, alienation and Bourdieu's (1990) related notion of hysteresis (cultural dissonance) important elements that may impact on citizenship as identity and as participation, producing new rights-claims and resistances to extant responsibilities. Citizenship clearly alters in action, experience and in terms of its legal envelope. This contingency is only partly, and tardily, reflected at the national level through legal definitions as action (unevenly) produces law and other responses that structure practice.

Prott (1998: 161), in another attempt at legal definition, leaves the door open for rights to be legitimate even if they are not enshrined in law or judicial pronouncement: 'A right is a claim that is enforceable within the legal system. It may have reached this status by being stipulated by legislation or by recognition in judicial practice or by tradition.' The legitimacy of rights-claims and of the tendency for rights to develop from below in the form of custom and 'tradition' is strong, as is that for rights-claims to emerge and then be adopted or enshrined at law. A recent case in point is where the courts accepted the moral argument in defence of (illegal) actions taken to damage and ground Hawk jet fighters bound for East Timor, Indonesia, where they would inevitably have been used repressively. This example also exhibits another important facet of citizenship that is discussed later: the development of globalised citizenship. There are other examples of this where different laws overlap conflictually, as was illustrated with certain provisions of the Criminal Justice and Public Order Act 1994 in restricting other welfare rights (see Weale, 1995; Parker, 1999a).

Batie (1984: 814) observes pertinently, in relation to land, that rights can also be viewed as any 'socially sanctioned activity on land', and Bromley argues that 'intelligible possession requires social sanction and social legitimation' (1991: 10). The argument is that the right/responsibility exists if it is observed – or perhaps observed sufficiently that rights flow from the collective and that the social contract should exist *prior* to the ownership (of land). Local rights are also possible: local laws are widespread, as are local common laws. It can be hypothesised that they may be made possible by local custom, culture or the conditioning and conditions of habitus and social field. Conversely, local law may have been introduced by elite groups to reflect dominant national constructions of legitimate responsibilities. Following Dahrendorf (1994), it could be argued that there are notionally concentric circles of rights, or at least, hierarchies of rights that operate across groups and across space, some of which are more 'embedded', difficult to repeal or better 'protected' by dominant groups (see Mann, 1987).

Legal rights as well as moral or customary rights are dependent upon prevailing conceptions of *legality* and similarly upon the individual's conception of *morality*. The legal framework is, however, rarely fixed or clear. It is also the case that the judiciary and the police interpret the legal widely (Bucke and James, 1998; Cooper, 1998). It is the acceptability of *claims* falling within the scope of the definition that requires further discussion. Keat and Urry (1982) make the point that the principles and rights of citizenship are presently confined and restricted to the operation of a distinct and limited set of political institutions. Therefore this 'political' state, which operates as a separate entity with a distinctive concept of citizens' rights *qua* human relationships, is a political statement. Morality is multiple, individualised, but any new claims have to pass through certain tests or obligatory passage points in what has been termed a 'blocked hegemony' (see Callon, 1986; Fudge and Glasbeek, 1992; Parker and Ravenscroft, 1999).

The concept of citizenship may be appropriated and shaped by different political philosophies. Marshall did accept that rights are sometimes utilised to reinforce existing inequalities (Marshall and Bottomore, 1992; Turner, 1986). But it is also clear that citizenship, as well as a bundle of clearly defined rights and responsibilities, may become a vehicle for ideology. Citizenship as activity may have a control function and exclude as well as liberate. The state can deploy citizenship in a way that legitimates particular identities as well as discursively constructing and defining activities as 'other'. Gramsci (1971) argues that this situation protects the interests of dominant groups who control the political field as part of hegemonic practice. Positions of power are often maintained by rights transfers made as concessions to maintain a hegemony. Clegg (1989: 160), following Gramsci, argues that hegemony involves the successful mobilisation and reproduction of the active consent of dominated groups as well as protecting the interests of the dominant groups. Popular interests and demands are incorporated where it is considered possible and support is organised for 'national', or perhaps now transnational, goals that serve the fundamental long-term interests of the dominant group. In terms of citizenship this is important: the construction of citizenship by the state and other interests represents a reservoir of brokered power. Hegemony is a process, and rather than accept the binary of dominant and dominated, this notion of relational, contingent (but stabilised) power relations is preferred (see Clegg, 1989; Woods, 1998a) and chimes with contingent citizenship.

Therefore, to maintain the social contract, an ongoing process of rights transfers to and from the public domain takes place. Some rights become firmly embedded in the culture of society while groups or class factions defend others less vigorously. Rights can be, and are, effectively used or enabled in the face of powerful, subjectively unjust opposition. In effect, moral citizenship can be deployed in the face of other practices or the exercise of power. One of the paradoxes of rights is, however, that they can be marshalled in order to marginalise less powerful groups as well as protect them. This is a key facet of the inclusivity/exclusivity debate surrounding rights and citizenship (see Giddens, 1998; Van Gunsteren, 1998).

Since the early 1990s, responsibilities have been prima facie devolved more to local scales of governance in an attempt to empower local communities (DoE/MAFF, 1995; DETR, 1998). Contestation and definition of citizenship rights and citizenship action politically are discussed in more detail in Chapter 3, but it is clear that as authorisation flows back from the centre or is empowered from transnational sources (e.g. the European Union, international campaigns and multinational corporations), opportunities for competing forms of cultural contract and citizenship may develop further (this began almost immediately in Scotland and Wales once the new devolved powers were granted; Bishop and Flynn, 1999; Scottish Executive, 1998, 1999). In one sense, then, localised citizenships are being encouraged and, linking back to the rural studies field, Murdoch (1999: 10) notes how rural policy-makers have been encouraged by government to allow a diversity of ideas and innovations to develop. It remains to be seen how such rhetoric will be realised.

Land, particularly rural land in private ownership, is a case in point here and is thematised in the examples discussed throughout the text. This illustrates how certain rights are viewed, constructed and treated as unassailable even when they are patently not so. It is also argued that the planning system has become a key element in the maintenance of rights distributions. The control of space is also a control upon citizenship even though authors such as Healey (1997) prefer to move planning theory onwards to concentrate on the role of space (and its prioritisation) rather than extending rights – a position that may be reconcilable with inclusive and deliberative citizenship. However, if planning is involved with the governance of space, it is also concerned with the governance of people. One might argue therefore that it has a crucial role in mediating part of the relationship between land and people and between different property right holders and others. Planning is subject to contestation and continual attempts to subvert and appropriate its aims as part of a strategy of (re)territorialisation of space by particular (largely dominant) groups. However, legal statements such as the European Convention on Human Rights or the US Bill of Rights, and more recently the 1998 Human Rights Act in the UK, theoretically allow for a range of minority interests to expose majoritarianism or partiality in challenges to policy processes and decisions.

Citizenship 'envelopes' and the cultural contract

A wider conceptualisation of citizenship can also take into account negotiative states and the notion of the cultural contract as introduced above. Different theorists argue over what constitutes (or should legitimately constitute) the social contract and the cultural contract. The state and elected governments attempt to impose regulation and uniformity in terms of rights, and impose aspects of national identity and national norms of behaviour. Van Gunsteren (1994, 1998) argues that every government action or, indeed, inaction can (and perhaps should) be regarded in the light of an impact on citizenship. As a corollary, citizenship envelopes are in a constant state of flux, developing and challenging norms and social fields. Such contingent definitions are also influenced from below at the level of the individual, group, community or interest. In this sense, citizenship is always shifting: it has become more complex, and the extent of rights and responsibilities, and the way they are mediated, are subject to prolonged political debate.

Different political positions will seek to define different activities or phenomena as being within or outside the citizenship envelope (see Sibley, 1992, 1995). That is to suggest that rights and responsibilities (both *de jure* and *de facto*) fall only to particular individuals and groups of individuals. The link between the varying conceptions of citizenship and the cognition and internalisation of *de facto* rights (in particular) as legitimate is important.

Few texts on citizenship have taken their analysis beyond abstract theorisation and applied citizenship in a specific context, or provided examples of policy and action within a citizenship framework, although authors such as Cooper (1998) have begun to publish applied research and Isin and Wood (1999) have recently considered cultural aspects of citizenship. A break from traditional accounts of citizenship, which have been rooted in political studies and state theory, is made here to consider 'citizenship' more broadly in order to incorporate elements of identity construction and person/selfhood as well as the more usual political/legal relationship between state and individual, applied to the rural context. This construction of citizenship explicitly allows for global factors and local features to be examined in terms of their effect on citizenship both as a set of rights and responsibilities and as participation in society at varying levels (e.g. local, 'community', regional, national, European, global).

Many of these are influenced by perceived and actual impacts of globalisation. Such 'new' citizenships include the notion of *cultural* citizenship, which has two elements. In this vein, Turner (1994) calls for a wider consideration of citizenship to extend the consideration of rights to include cultural citizenship: first, the defence of particular aspects of cultural identity and history, which prompts demands to preserve or reinstate traditions and customs; and second, for these to be reinforced through institutional support, e.g. state provision of Welsh-language teaching in Wales, or indigenous North Americans to be allowed to pursue old claims over land through modern legal institutions (see Jacobs, 1998).

Cultural citizenship can be said to be part of a postmodern view of citizenship where status, identity and participation are fluid, and recognised as such, for many individuals – both advantageously and detrimentally. In this view, citizenship becomes more about practice and particular activities and distinctive actions – that is, from being to doing (both strategically and in terms of the 'everyday'; see Shotter, 1993). Cultural citizenship, as with 'traditional' citizenship, can be viewed as a citizenship form intended to protect existing features of social and economic life in a given area, or for a particular group. It can also be viewed more expansively: to infer a citizenship based on progressive, pluralist 'becomings'. Changes in culture, societies and economic structure fed the growing interest in citizenship during the 1990s, and many other facets of citizenship have been auspicated. The possibilities for new global, often group interest or non-human-based rights have also been claimed (Isin and Wood, 1999).

The second aspect concerns the emergence of citizenship(s) that reflect cultural difference or reflect an internationalistion of culture. Both are linked to globalisation and the development of postmodern society: 'very few modern societies have such cultural uniformity. Multiculturalism is an inevitable consequence of globalisation. Finally, there is the view that formal participation in the national culture may simply disguise major *de facto* forms of exclusion' (Turner, 1994: 159–160). Thus far, accounts of cultural citizenship have limited themselves to notions of groups claiming, as of right, that certain practices or material elements of their locality and culture should be preserved and protected by the state or by supra-state authorities (see Cooper, 1993, 1998; Prott, 1998). The notion of cultural citizenship can be interrogated by asking, as Turner (1994) implies: what of cultural citizenship in terms of action and of participation? And, further, what of the cultural arbitrary imposed on others through (in)action (cf. Bourdieu, 1990 on symbolic violence; also Parker, 1999b). This surely lies at the heart of post-citizenship analysis, in terms of both action and non-action: of engagement and of apathy and in terms of the active construction of citizenship by the state. It is clear that reorganisations of space/place and identity are likely to produce new and complex communities and communities of interest/identity. It should be recalled, from earlier discussion, that regardless of such changes, citizenship remains inherently exclusive. The key difference in the future might be explicit recognition of contingency and mutual respect for difference.

'Good' citizenship also implies an element of knowing or social dexterity: of competency. In the rural context, Cloke and Little (1997) refer to the notion of cultural competence whereby people require certain knowledges in order to function at the level of the community. By contrast, Thrift (1983) back in the early 1980s identified at least five types of 'unknowing' that can be applied here. 'Unknowing' is linked to the notion of 'becoming' in challenging the implied exactitude of 'competence' – the implication being some form of 'mastery' of the locale, or at least the suggestion that relations and knowledges remain stable such that knowledge/culture/practice and other elements of cultural and symbolic capital remain valid as 'cultural tokens' (Bauman, 1992; Bourdieu, 1990; Fine, 2001). This carries with it an implicit assumption of stability or constancy.

How the cultural citizenship concept itself is understood and how it may be relevant in the rural context may be reworked. The notion of a cultural citizenship ties notions of competence with symbolic exchange relations (Baudrillard, 1981). Aspects of performativity, counter-cultural shifts and resistance to top-down legal binaries such as right/wrong, competence/incompetence are also implied. What we may see in terms of the post-rural condition and way of seeing (Murdoch and Pratt, 1993; Berger, 1973) are contestations over definitions of rurality, as exemplified in the Countryside Rally in 1997 and the Countryside March in March 1998, which are explored in Chapter 7 (see also Woods, 1998b; Monbiot, 1998; Norton,

2000). They may be read as attempts to resist socio-economic and definitional change – from within, about, above and below. The 'competence'
dimension is not stable, however, and different contexts may require
different skills or competencies. Each culture is subject to challenge and
modification both locally and from afar, through social recomposition, technological advance and economic change. Such competencies also shift with
the requirements of cultural and political citizenship.

Globalisation and the fragmentation of citizenship

It is important to situate citizenship within a global context, a scale of
relations that has been increasingly subject to rapid economic and political change, particularly in the past ten years or so. The role of the media
and technologies of information and communication have been important influences. Increasingly, the spectre of 'public opinion' is more often
the opinion of newspaper editors and television producers, and thus the
role of the media in forming concepts of citizenship is also mentioned in Chapter 7, but this too is a globalising influence. Globalisation is
a topic that has already generated prolonged and extensive debate (e.g.
Bauman, 1998; Beck, 1998; Ray, 1999; Urry, 2000) and therefore one
whose complexities and nuances cannot be adequately examined here.
Giddens (1998) highlights the changes in economic structure and
technological advance that have led to what Lash and Urry (1994) term
reflexive modernisation and to Beck's notion of 'Risk Society' (Beck, 1992,
1998).

It has been traditionally argued that citizenship and citizen action is
largely constructed and orchestrated by the state (Held, 1989, 1995). This
view has altered during the past ten years or so, with increasing attention
being paid to the influence of the global (and the local) on citizenship
and the nation-state. This has led to the popularisation of the term 'glocalisation' to denote the increasing significance of global culture and economic
forces and the response of localities and regions – often occluding the
nation-state as part of this process (Robertson, 1995). Urry (2000) asserts
that there is a state of global disequilibrium and of mobility that impacts
on the ability of national governments to know and fulfil their part of a
social contract, let alone a wider and complex cultural contract: 'Just at
the moment that everyone is looking to be a citizen of a society, so global
networks and flows appear to undermine what it is to be a national
citizen . . . contemporary citizenship can be described as post-modern'
(Urry, 2000: 162). Thus, in Urry's view, citizens look for alternative
ways of furthering their interests and for alternative identities. This involves
a shift from the traditional – that is, from the pursuit of social rights
and valorised citizenship status. Rights-claims are one method of attempting
to gain and broker self-interest, but may be portrayed and lodged as
part of a group interest. Urry refers to the development of a post-

modern citizenship as involving the interplay of the global and the local in a symbiotic relationship. He also plays up the potential importance of locality, where small-scale change at the local level can have 'unpredictable and chaotic' consequences for the bigger scales of governance (Urry 2000: 210). Hence the re-imagining of citizenship has important spatial dimensions.

When one examines citizenship alongside its Marshallian definition it is clear that there is not equality in terms of rights and duties, liberties and constraints, and powers and responsibilities. The fragmentation of contemporary culture means that any statist construction of citizenship would have difficulty in satiating all citizen rights-claims. The liberal construction of citizenship is one that attempts, in opposition to social and cultural trends, to define *narrowly* what constitutes acceptable, proper or 'good' citizens' behaviour (Ravenscroft, 1993). The varying constructions of citizenship envelopes can be narrow or wide, and indeed both are dependent on the behaviour of the citizen and the conditionality of the citizenship paradigm. A useful and important distinction here is between an *assimilatory* citizenship ideal and an *integrative* citizenship concept (see George and Wilding, 1985). In the former, citizens are forced to adapt to a single cultural formula, while the latter allows a broader and diverse cultural milieu bonded by common themes or 'codes' (Van Gunsteren, 1994).

While rights and responsibilities and their mediation are important in understanding changes in social and political terms, other powerful forces and interests such as big business, local action groups or cultural idiosyncrasies influence contingent distributions of entitlements and obligations. Alterations in the distribution of rights can be viewed in all policy to some degree or other. Therefore, it is not only policy and practice which claim overtly to involve citizen empowerment that can be discussed within the framework of citizenship theory.

Citizenship in the postmodern era

The contents of citizenships are variable from place to place and understood differently from person to person. They are recognised to differing degrees by powerful agents, individuals, sociations, local elites, the local state, and the nation-state and beyond. It is the way in which power, through the imposition and receipt of 'citizenship' and how individuals react as 'citizen', that is a central concern here. Urry (2000) has commented on the numerous labels that have been generated in recent years to distinguish supposedly new elements of citizenship. Examples include the *ecological* citizen (Van Steenbergen, 1994) and the *environmental* citizen (Newby, 1996) whereby rights and responsibilities towards the environment, wildlife and the ecosystem are said to have become more important, particularly in the light of more stress being placed on the importance of responsibility in UK politics (see Taylor, 1995; Hutton, 1997). Of course, many such global responsibilities (with associated rights) develop from localities or communities of interest and are resisted more by extra-national bodies such as big businesses than by nation-states themselves.

Increasingly, Heater's point about centrifugal forces that threaten the established parameters of citizenship is prescient: there are a host of other 'citizenships' being propounded (Van Steenbergen, 1994; Bulmer and Rees, 1996; Isin and Wood, 1999). This has led to even more recent pronouncements about the death of citizenship (Falk, 2000), much as, for example, town and country planning was said to have expired in the 1980s (Ambrose, 1986). Instead, multiplicities of citizenships are evolving to form a fragmentary *bricolage*, again similar to the much-touted fragmentation of planning over the past fifteen years (see Brindley *et al.*, 1989, 1996). A good illustrative example of the type of manifestation of cultural or postmodern citizenship can be read in the way that some members of the UK's ethnic minority populations feel allegiance to cricket, football or rugby teams representing the countries of their familial origin rather than their current residence. Perhaps such developments require redefinition and a wider perspective on global and localised conditions rather than premature certification.

The rural context is interesting, as it is claimed to be 'fluid space' both in terms of cultural change, and also in terms of the differential degree of regulation that has been achieved as compared to urban space in some areas. This practical liminality remains problematic for the state to keep under surveillance and regulate effectively. Much of the dissonance implied is silent or covert, however (Parker, 1999b), but there are important examples to be highlighted where the countryside and groups with interests in the rural have overtly and effectively maintained differentials. A classic example is the successful evasion of much of the regulatory framework of the land-use planning system on the part of the farming lobby (see Bishop, 1998; Cullingworth and Nadin, 1997; Gilg, 1996; Winter, 1996; Cherry and Rogers, 1996; Curry, 1994), which arguably led to the destruction of numerous environmental features and the construction of many unsightly agricultural buildings (Shoard, 1980, 1987; Bowers and Cheshire, 1983).

As a result of increasing reflexivity and mediation, more people are less unaware of the conditions of their existence and of the conditions of others. Part of this process, it is argued here, lies in the broadening of citizenship and the need for citizens to exercise political agency in novel and informed ways. The development of new forms of citizenship are at once enabled and undermined by reflexive modernisation. Numerous labels have been developed to reflect changes in the dynamics of rights and responsibility relationships. Much talk has been about the development of a *European* citizenship (Institute of Citizenship, 1992; Institute for Citizenship Studies, 1999) whereby formal rights and responsibilities are created and guaranteed at the level of the European Union. Similarly the idea of *global* citizenship has gained currency, increasingly as the realisation of global responsibilities and cultural rights is divined (see Falk, 1994; Turner, 1994).

(Re)spatialising citizenship

Definition of 'citizenship' in a spatio-temporal sense is made complex by the heterogeneity of individuals (see Passerin-d'Entreves, 1994), let alone the relationship between that individual and his/her institutional relationships or the interests and associations that the individual (or group) may develop. Such shifts may engender the (re)evolution of some 'rights' of citizenship (see Cooper, 1993; Dahrendorf, 1994; and Chapters 3 and 4 of this book). This aspect of citizenship presents considerable challenges for the state and other actors to oversee, regulate and accommodate. Historically it has been argued that public space was the only space that was political space (Kearns, 1995) and that private space was subject to private, domestic control. As will be examined later, this perhaps helps us to understand more about struggles over the control of rural land. Access to the land and controls over its use erode the private governance and territorialisation of that resource.

There has been a growing literature in cultural geography on the topic of cultural politics and space (see, for example, Keith and Pile, 1993). There have been at least three wide concerns explored therein: the politics of place, the spatialised politics of identity and the significance of space for local governance (Painter and Philo, 1995; Marston, 1995). All are important here, largely because they involve struggles to define and control space and place, for both object and subject, in terms (as with citizenship) of status, identity and activity. Spatiality and territoriality represent the inextricable interrelationship of space and society, and by association the idea of citizenship construction. Keith and Pile (1993) also suggest that there are three key areas for a politics of place and identity: locations of struggle, communities of resistance and political spaces. Through the ensuing chapters all three aspects, as they relate to the countryside and land, are touched upon in the examples of citizen action and policy for and in rural areas and communities.

Space has been recognised by citizenship theorists as important, and conversely geographers have become interested in citizenship (see, for example, Smith, 1989, 1995; Fyfe, 1995; Blomley, 1994). Isin and Wood (1999) argue that space is an important constituent element for a group's identity and its capacity to claim rights – again both on a practical level and on a level of the symbolic. Space provides the material platform for the expression of identity (as with the body and the concept of hexis developed by Bourdieu) and for group expression and the expression of power (see Crouch and Matless, 1996; Radcliffe, 1993; Bourdieu, 1977, 1990). In this sense, the land and the countryside is used symbolically by groups such as landowners and by farmers not only in terms of its functionality but also in terms of its 'meaning'. To illustrate this point, one such discursive strategy used by these groups may be labelled the 'discourse of stewardship' (Parker and Ravenscroft, 1999), whereby space and its control are claimed on moral, environmental and historical grounds as much as on legal, property-based rights grounds. Multiple interests make this type of association – partly as a means of legitimising rights-claims. For example, one claim is that the land has been and should be held in the benevolent and unproblematic control of its stewards – a claim that has come under increasing challenge over the past fifteen years or so. Woods (1997) argues that this view has been replaced by a discourse of 'community'; however, in terms of land particularly, it is argued that there remains a clear stewardship rhetoric (see, for example, Country Landowners Association, 1998). The well-financed Countryside Alliance, in its careful manipulation of the media, has indeed attempted to suggest that it represents the (true) rural 'community', as is explored in Chapter 7.

As an example that is further unpacked later, it is possible that much of the resistance to the use and existence of rights of way on the part of landowners/occupiers has been related to issues of identity and control. The presence of 'activities out of place' (see Cresswell, 1996) perhaps threatens the integrity of the farmers' identity. Rights of way become threats to the integrity of the spatial logic of power. Similarly, user groups and other protest groups claim rights over space (including blocked or disputed rights of way) in much the same way. They exercise physical occupations of space – for example, by mass trespass, squatting and demonstrations at symbolic locations (see McKay, 1998). The examples used in the latter chapters draw on such competing claims and definitions of rights *qua* activities and participatory styles to deepen the consideration of this element of citizenship.

Citizenship construction in terms of the UK countryside has in the past exhibited characteristics of paternalism, rather than of a narrower liberal citizenship; such characteristics have been linked to the historical development of land and rights distributions. There have been varying degrees of benevolence on the part of the state and private landowners in terms of rights and responsibilities (Parker and Ravenscroft, 1999), and increasing agglomeration and rationalisation of power – crucially in terms of land and land use.

Contesting citizenship in the countryside

'Rural studies' is populated with researchers from across the social sciences who draw on the rural for a host of reasons (see, for example, the exchange between Miller, 1996; Cloke, 1996; Crow, 1996; and Winter, 1996). This calls into question whether there is such a neat rural field of study at all. Disciplines such as agriculture, geography, planning and leisure studies have continued to be strongly represented in terms of countryside research agendas. Those more traditional or fixed lines of research inquiry are increasingly reinforced and sometimes challenged by other, newer work that aims to investigate aspects of life in and impacting on the countryside. Different conceptual and policy-relevant strands that have traditionally been debated in the abstract link various matters of concern in rural studies. It is timely to apply such areas of theory explicitly to the rural and also to tie in our investigation with government attitudes to both the countryside and towards citizenship construction.

Here, the rural is examined primarily from the perspective of an arena within which wider issues of identity, socialisation and relations of power in a differentiated, multiformly defined society can be explored. The concept of the post-rural (see Murdoch and Pratt, 1993, 1994) is used as an important contemporary perspective that bears on the historical examination of rights, power and citizenship in the countryside. The term 'post-rural' can be understood in a variety of ways. The concept is related here to the way in which rural space has become understood as being more diverse (and often obscuring a range of inequalities). It is also imagined space, interconnected with urban and other places through diverse network relations. Indeed, this reflects the concern among many academics engaged in rural studies to research and assess the rural in a manner that uncovers 'otherness' and examines the plurality and diversity of rural place, space and agency (see Cloke and Little, 1997; Cloke *et al.*, 1995).

In the case of access to the countryside (for recreational purposes), there is clear resonance with debates over citizenship, space and land use. Access is one of the clearest indicators of how use of the countryside and imagined usability of the countryside have changed (or are seen to have changed) from the period when rural land was first and foremost a productive space to an arguably post-productivist era of the rural as amenity or consumption space. By and large, the reaction of the landowners as stewards has been to argue that any use of land must imply commodification and requires payment. It also indicates in a clearly spatial way how citizenship operates as an exclusionary device to bar entry to material space as well as conceptual openings. One aspect that Marshall envisaged was that citizenship could indeed preserve class inequality as much as it could eradicate other aspects of inequality (Marshall and Bottomore, 1992).

Recent changes in the regulation of the countryside have brought widespread debate about issues such as commodification, counter-urbanisation and increased demand for development. All those issues impact on the habitus and the construction of citizenship in the countryside. There are a multiplicity of combinations of rights and responsibilities that could be legitimated at any particular time or in any particular place. The consideration of land ownership and rights as responsibilities imposed from above in relation to land forms the core of the following section. Land as a form of capital, and the power accruing to the holders of these assets, impact on citizenship in numerous ways.

Countryside matters have become the concern of a wider range of groups and individuals and there has been a spate of high-profile issues prompting a reconsideration of attitudes and policy towards the countryside. These include agricultural restructuring and support, food safety, appropriateness of planning regimes, environmental protection issues, leisure pursuits and their legitimacy and impacts (including the blood sports debate), and issues of social exclusion, both from and in the rural. Such factors are important here; specific aspects of countryside change form the context against which the arguments over citizenship are discussed. Indeed, each of the areas mentioned above has impact on, or is influenced by, legal and cultural rights and responsibilities, and all can be influenced by citizen action. For example, it is argued that the changes in terms of social/class composition in many areas of the UK has allowed the rural to become a more 'fluid space' where a range of different outlooks and activities can be better expressed (Savage *et al.*, 1992; Murdoch and Pratt, 1994, 1997; Clark *et al.*, 1994). In the case of the countryside, it is possible to discern that the legitimacy of rights is affected by social (re)composition in the countryside. The 'social sanction', mentioned above, to enforce legal rights over other(ed) rights may rest with a particular group in any particular locale, rather than the community as a whole, dictating the legitimacy of other(ed) rights, and this can create intra-community friction. In the past, perhaps this power of intimidation may have rested with the 'squirearchy' (Newby, 1987). Over recent years the balance has been swinging towards middle-class incomers (see Marsden *et al.*, 1993; Thrift, 1996) who have been recently labelled the 'New Magistracy' (Stewart and Stoker, 1995; Murdoch and Marsden, 1994).

Rights are, however, symbolic as well as functional and are reflective generally of the times in which they exist or are exercised. This is another element of the cultural specificity of rights when culture is measured in time as well as through activity. This is reflected in law, where, for example, a right of way is presumed to exist if twenty years' uninterrupted use is proven (Riddall and Trevelyan, 1992). Contemporarily it is argued that rights are similarly reflective of time, space and agency. Within localities, rights, which may be customary (the *lex loci*, which may indeed be state enshrined), may to some degree be discordant in relation to other spaces in terms of regulation. Such a thesis relates neatly with the notion of Bourdieu's concepts of habitus and field (Bourdieu, 1977, 1990) and the associated idea that locales (Giddens, 1979; Peet, 1998) exhibit cultural identities and practices which may render them in some ways autonomous, self-regulating, or more likely parallel to national regimes.

The emphasis that is placed on the different sides of the citizenship equation is important in terms of day-to-day politics and issues concerning political accountability, and empowerment. The construction of citizenship is infused with the political priorities of the elected government, and therefore to some extent the citizenship envelope moves with that worldview. Liberal conceptions emphasise obligations, duties and responsibilities, while traditionally the social democratic left have concentrated on provisions, entitlements and rights.

The land and the citizen: citizenship rights and private property rights

Increasingly, citizenship may be seen as what one does across a range of measurement scales. Attention is given here to the relationships between private property rights and the rights of the citizen: between private and public rights (and between *de facto* and *de jure* rights). There are tensions where private rights infringe on other 'freedoms' (and vice versa), and the attitude of the state in respect of the distribution and enforcement of such rights-claims is important. The treatment of various claim-rights by the state and by powerful individuals is importantin constructing and maintaining a dominant conception of citizen rights and responsibilities (see Vincent and Plant, 1984; Dahrendorf, 1994; Marston and Staeheli, 1994). Therefore, how individuals behave, and particularlyhere how they behave and use land, is important. They challenge territoriality and implicitly challenge (or reinforce) extant rights/responsibility distributions.

Marxian philosophers have made much of the conflict between the rights to liberty and those of property (Keat and Urry, 1982; Becker, 1977). Turner (1986) states that one of the roles of bourgeois freedoms is to uphold the right to own property, and Held (1989) asserts that this right was one of the first civil rights entrenched into Britain's developing liberal civil society. The delimitation of rights and responsibilities of the citizen in relation to land and its administration is an area of social science that draws attention from theorists from across the political and disciplinary spectrum. There has been an increasing amount of interest in the re-examination of the structure of private property rights, especially in relation to citizenship rights. Newby *et al.* (1978: 345) state that:

> We believe the rights associated with property to be such a taken-for-granted (and hence hegemonic) aspect of the social structure and to be fundamental in both shaping the system of rural social stratification and prompting a good deal of the political activity in which farmers and landowners engage.

More recently a reminder of the importance of property rights in the rural context was issued by Marsden *et al.* (1993: 69):

> The ownership of rural land may be of modest significance but the local distribution of property rights remains crucial to the pattern and processes of rural development. This is pre-eminently because of the continued association between control over property rights, local elites and the rural class structure, and the focus upon land as the means of realising many public policy objectives.

The exploration of property rights and citizens rights in the context of countryside policy is one of the crucial steps leading to an understanding of the effects of such policies. Such an analysis inevitably highlights the failures of policy aimed at social and economic change (see Cullingworth, 1994; Healey, 1997). The failure of progressive land policy (such as the Land Compensation Act 1967 and the Community Land Act 1975, both since repealed) is due, in no small part, to the entrenched claim-rights which landowners exercise in order to maintain a status quo. It is this fundamental obstacle which planners and policy implementers face. The 'dominant' ideology in the UK has supported the status quo with regard to property rights (see Abercrombie *et al.* (1980) for a fuller, subtler exposition of the development of property rights and ideology, and Marsden *et al.* (1993) for research examples).

There was a short period following the Second World War when the political climate enabled the Labour government of the time to bring certain property rights under state control, and many of the changes brought about then persist. The use of power by the state in respect of those rights has meant, however, that they have been colonised by the property-owning classes to maintain amenity and land values. Mediating the establishment of certain rights at the expense of other acquired rights held is problematic in so far as property rights generally are firmly entrenched as citizenship rights. Becker (1977: 112) notes that:

> The sorts of property rights which can be justified vary with social circumstances. . . . Thus rights obtained justifiably in one time and place and perpetrated by justifiable transfers . . . may turn out to be unjustifiable in terms of a good distribution for the current social situation.

The maintenance of economic interests is one of the main points of political contention when seeking to adopt social citizenship rights. In terms of property rights there has been extensive discussion concerning the legitimacy of present rights distributions, not least in terms of how to realise social (and spatial) rights-claims in the face of private property rights.

Power and property rights

The analysis of power in this context is the most important feature of ownership (Denman, 1978). The landowner in most respects holds the power in land, and has the power to waive certain rights. Notable erosions of private rights have come about through progressive planning legislation such as the 1947 Town and Country Planning Act. Thus trades in rights are trades in power. Hohfeld (1919) regards the existence of a power-right as where a right-holder may (whether morally or legally) alter rights, duties, liberties or powers of another. It is debatable whether power-rights in this context will involve two-way power brokerage between the landowner and the state. Power relations are crucial: the liberal right believe that 'power' should rest with the individual while the social democratic left have traditionally considered that such power rights should be vested in the state on behalf of the population as a whole. In both instances the power is enforced by the state (local, national and supra-national).

It is clear that the intervention of the state, in particular through the planning system in the UK in respect of real property, is viewed by libertarians as unjustifiable, while social democratic theorists view intervention and nationalisation of some property rights as necessary in order to control inequalities. One of the main divergences of opinion (after Plant, 1994) therefore concerns the 'sanctity' of private property and the ongoing issue of what rights and responsibilities should accrue to whom and in what circumstances – and, additionally, where and how social, cultural and customary rights (including nascent rights) are recognised and constituted. In the context of contemporary concerns over the environment and reassessment of land-use priorities, certain previously accepted rights distributions have come under increasing attack. There is a growing recognition of the social interest in land and an erosion of trust in traditional 'stewards' as well as an increasing concern for 'amenity' (Bromley, 1991; Cox *et al.*, 1990). Together these trends are powerful, and certain environmental rights and responsibilities are being continually contested and traded. Again, the planning system has been vaunted as potentially the best vehicle to act as intermediary between state (public) and the (private) landowner; even though it requires further democratisation and transparency for such a role to be taken on (see Selman, 2000; Healey, 1997; Allmendinger, 2001).

There appears to be a need to question further the basis for authority and control over land and its management. This is a process that has slowly and stutteringly been occurring through the 1980s and 1990s. Recently, for example, alterations and accommodations are being made in terms of access to the countryside for recreation, or for other rights (such as fox-hunting) to be removed. Such challenges are challenges as much to the cultural domain of the land manager as to the activity or land use itself. They are as much a challenge to the competencies and cultural rights of traditional land interests (and structures and processes of governance) as to claims for rights.

It is increasingly clear that the 'social glue' (Pahl, 1998) of employment, relative immobility, etc. that used to bind groups together in rural areas is being replaced with new aspects of self-interest (Cherry and Rogers, 1996). This process tends to reconstitute social relations in such a way that the spatial loses its primacy. However, one spatial outcome is the re-imagining of place and social field or standing conditions of rural areas. Extant power relations may be felt to apply less and less. One example is the ongoing conflicts between land use and ownership and the cultural right to amenity value and experience/consumption of landscape; another is the disruptive, often complex, cross-cutting interests of different people who may be located physically in the countryside. This leads to the reworking of citizenship in the countryside in terms of culture, identity and participation. As a result, traditional conceptualisations of what it means to be 'rural' hold little relevance for many rural-dwellers or those for whom the imagined rural is a far cry from the day-to-day 'reality' of their physical and social surroundings.

The foregoing changes of perspective are partly reflective of what Giddens has termed 'life politics', where people are increasingly reflexive and critical and incorporate global concerns and viewpoints as well as local and personal ones: 'political issues which flow from processes of self-actualisation in post-traditional contexts, where globalising influences intrude deeply into the reflexive project of the self, and conversely where processes of self-realisation influence global strategies' (Giddens, 1991: 214). It is also key to this text that such changes in culture and politics may lead to a re-examination of citizenship by national-level governments, and may lead new expressions of citizenship and governance more generally to be recognised as such, especially in the rural context. This leads to a discussion here of how citizenship may be reconceptualised for a globalised, postmodern era; how government might seek to envision citizenship and rural regulation that is conducive both to existing rural people and to those wishing to use, or in some other way be involved with, rural areas (e.g. through conservation, or concern over sustainability or food quality issues). These considerations may play out in policy terms to require reorganisation and a reconfiguring of, for example, the land-use planning system, or the drastic rethinking of the basis and preconditions for land ownership and payments of state monies to the agricultural sector (see Ilbery, 1999; Fairlie, 1996). It has also given rise to restrictive measures, as discussed in later chapters, and has recently seen the UK government passing the Human Rights Act 1998, which 'guarantees' certain 'rights' for the first time. This represents a merging of responsibility of the judiciary and the legislature in the UK (Home Office, 2000; *Guardian*, 2000c). It is likely to alter the way that citizenship develops and alters across space and time, with particular implications for attempts to act in the group or 'national' interest through the planning system (see Johnston, 2000; Blackman, 2001).

Land claims, historically concerning production or benefit streams (see Bromley, 1991, 1998), are shifting towards consumption or different kinds of benefit. Since, from the Romantic period onwards, a desirable image of the rural has been constructed for public consumption, there have been claims over the rural as a consumption space and for consequent cultural rights over the land to match the claims of older date for land to be distributed fairly; or for land to be held as a commonwealth. In terms of the self, arguments in the economic sphere have led towards shifts in the cultural and the political. The dynamics of power alter, as do the dynamics of the imagination of space and consequent perceptions of legitimate action. It is more that citizens imagine and claim rights in and over the rural than actually exercise such rights.

Conclusion: towards fluid, post-national citizenships?

Since Marshall, the notion of the UK as a coherent and stable nation has come under increasing challenge from within, and as part of the development of a critical academic analysis of citizenship and of state formation. The types of cultural, technological and economic change that have been experienced, especially since the Second World War, have meant that traditional concepts of citizenship have failed to match with the diversity and aspirations of very many 'British' people. In short, assimilation is impractical and probably undesirable. Consequently, citizenship as a concept of the national has come under increasing strain. Citizenship in the postmodern, post-rural context requires different conceptualisations that allow and recognise contingency, conditionality and changes in power structures, including allowance for 'external' contributions to political and social life to be heard and incorporated. Increasing reflexivity of individuals, the environmental agenda and long-standing historical divisions that have existed in the UK all contribute towards feelings of frustration with old systems of governance and distributions of benefits streams (see Bromley, 1991).

In terms of the countryside, and land in particular, it is perhaps time to restate how the control of ownership and regulation of the land is an integral, practical and symbolic aspect of citizenship (both cultural and political). The way that land is managed is of itself part of the social contract, yet it has rarely if ever been expressed in this way. In much the same way, the idea that 'citizenship', as an arrangement between the state and the individual, is a social contract has now found currency in popular political rhetoric. It is perhaps time specifically to examine land (its use and governance) and citizenship in a similar fashion.

All this means that the role of the state in terms of nation-building and citizenship is changing. Even if government and state still attempt to act as though they can control culture, identity and citizen/consumer action domestically (and even the actions of those at a distance who impact on the nation-state and nation space), the balance is shifting such that direct and participative democracy (albeit via a minority) is supplanting more traditional, modernist structures. The state, politicians and dominant groups are thus faced with an increasingly self-preservationist desire to hold together at the national level. This may be operationalised through constructions of national identity, exhortations to cultural affinities and the use of rhetorical devices and regulatory discourse such as the Criminal Justice and Public Order Act 1994 and the recent Terrorism Act (Home Office, 2000). This is done in the face of erosive influences on the national from the local and the global and from (networked) communities of interest rather than national or spatially specific communities.

It is clear that any rights distribution is a contingent distribution and therefore the legitimacy of such rights may be temporary, if not continually contested. Some rights at particular points in time are viewed as unassailable – for instance, the right to own property in civil terms, the right to vote in political terms and the right to an education in social terms. Various critiques of citizenship theory have outlined the dynamic nature of rights and the processes by which rights are won, lost, maintained and reinforced (see Held, 1989; Heater, 1990). There is interdependency between different sets of rights, their construction, and the political impacts of those constructions. They are also, on closer inspection, complex sets of rights, responsibilities, entitlements and obligations. Postmodern citizenship extends far beyond the definitional envelopes provided by the state, and in political projects every action can be regarded as political and as part of citizenship as process. The construction of production and consumption rights and their meanings are substantially formed by the processes taking place in the social world (Mouffe, 1993; Clark *et al.*, 1994). The effects of specific policy and economic restructuring on production/consumption rights (particularly in terms of the interface of property rights and citizenship rights) are such that claims to vary existing rights distributions are likely to develop, given, for example, the economic position of agriculture and the consequent need for alternative land uses and incomes from the land and the countryside.

In Chapter 3 the attempts of consecutive governments in the UK to 'manage' and construct citizenship are detailed. In this and later chapters we examine how those projects have been designed in terms of the rural or how those projects have affected citizenship roles for people either living in or interested in some way in the countryside. This is done in the context of government and nation-state attempts to shore up nationhood and responsibility using a modernist model that appears to have become less and less relevant in a globalised or 'global village' context (McLuhan and Powers, 1992). Chapter 3 provides an historical review of land and rights. Then contemporary examples of how citizenship policy, rhetoric and citizen action are currently being played out in (post)-rural contexts are examined.

3 UK politics and the citizenship debate

Introduction

This formally political segment contextualises and links the theoretical frame of citizenship to the historical and contemporary application to land and people explored in Chapters 2 and 4. Consideration is also given to the underlying political theory that legitimates or challenges such citizenship constructions in the light of the examination of citizenship theory in Chapter 2. The central concerns of this chapter are the contemporary policies and attitudes of successive UK governments, which are explored in terms of citizenship, the use of citizenship rhetoric and the claiming of rights and enforcement of responsibilities/obligations. Particular emphasis is placed on the political project of Conservative governments in the UK between 1980 and 1997, which invoked the concept of 'citizenship', and the similar deployment of communitarian citizenship rhetoric by 'New Labour' under Tony Blair since 1997. In all cases these governments have designed policies to construct citizenship and have used rhetoric of 'active' or 'engaged' citizenship. They have promoted a new governance based on an explicit conceptualisation of citizens as consumers and superficially as part of 'communities'. This is critiqued and linked to the changing and claimed relations for citizens (and consumer-citizens) to the land.

Citizenship is potentially a political formula by which interests (be they class-based, minority, group-based, radical, etc.) and the claims derived from such interests can be heard and resultant issues contested freely. In the context of UK politics a more inclusionary, flexible and participatory citizenship would be welcomed by a number of factions, but at the same time such a citizenship could similarly compromise the existing political arrangements, which are currently stabilised through the system of representative democracy. As detailed in this and later chapters, small groups of people are assembling and acting on a range of specific interests as part of reflexive projects of self-realisation. Some are breaking free from preoccupations with the national or domestic to incorporate considerations of wider 'life politics' and increasingly to consider global affairs (see Franklin, 1998; Bauman, 1998). More widely – and further research is certainly required – people are finding alternative means of expressing political views, for example through consumption and related practices.

In geographical terms citizenship may be becoming less about status tied to territory, and more about network membership and the holding of particular reserves of social/cultural capital. The focus of this book is on the countryside, and it is discerned that countryside issues can arouse deeply held views and beliefs (Cherry and Rogers, 1996) and raise competing individual conceptions of optimal citizenship, as well as illustrating state (and alternative) constructions of the rural. The large-scale social and economic changes impacting on the countryside over the past thirty years have not yet prompted government to seek a radical plan for rural areas or the use of rural land. In particular, the way that land has been used and regulated has been subject to trade-offs over time, and this incremental process continues. The recent political history of citizenship in terms of formal politics and shifts at the global level have forced a persistent use of citizenship to reinforce the 'national' and to encourage greater social efficiency over the past ten years. Attempts on the part of John Major and Tony Blair, as UK prime ministers, have been made to utilise 'citizenship as doing' in order to carry through programmes designed to stem the flow of state expenditure and encourage active 'self-help'. The way in which the citizenship ideal has been appropriated and challenged in this way is discussed below.

Society, the state and citizenship

This chapter has several strands. First, it examines how successive governments in the UK since 1979 have attempted to frame, construct and operationalise citizenship and how those attempts have been played out in a rural context. Second, it looks at how groups and individuals have responded to such exhortations and how alternative 'citizenships' have begun to emerge, especially in relation to struggles over rural space, place and identity. Finally, it looks at how traditional citizenship has required division and polarity, as a reflection of the binary of the legal system. Often groups have competing claims that may require different definitions or envelopes of state-sponsored citizenship, and dealing with those claims may require the legitimisation of exclusionary measures aimed at certain groups. These themes are explored further, using examples, in subsequent chapters.

Recently, political commentators have begun to recognise that the idea of citizenship could have a more multidimensional nature, or at least a more nuanced and complex constitution, than the tripartite evolutionarist perspective adopted by Marshall in the 1950s (Marshall and Bottomore, 1992). A reinvigorated citizenship that challenges dominant constructs and limiting definitions can be viewed in terms of both the new types of rights and responsibilities that might be added to the citizen envelope, and the enablement of new dimensions of citizenship identity and practice – that is, as participant in civil society rather than one who receives status and contract in a preformed and implicitly immutable package. In short, citizenship should enable and be enabling; be performing and performed.

Inevitably, as citizenship alters and new rights are claimed (and when citizenship is reconceptualised), there will be conflicts with other 'rights' held and valued by particular groups. At these kinds of friction-points the types of rights and obligations that are imagined, local or cultural can be located. One outcome of the contestation of rights is that citizenship debate tends to lead to adversarial contests. Such manoeuvrings require coalitions of similar interests to align themselves, or for powerful interests to appropriate agendas and claims as part of a strategy of 'bundling' or enrolling claims. In this process competing 'citizens' are required either to undermine other claims or to portray their own claim as homogeneous, or as complementary to dominant rights distributions.

It follows that citizenship and the citizenship envelope operate as much as a tool of exclusion as a vehicle for inclusionary claim. This is a common criticism levelled at citizenship. By its very nature, being a citizen and accepting the legitimacy of certain rights is exclusionary. Arendt (1977) sees this as an agonistic model. It has operated in this way, and still does so, as an important part of the project of modernism. One of the key features of postmodernity or late modernity, however, is the increasing awareness of contingency and, it follows therefore, of the contingency of rights and obligations. Citizenship can be seen as fluid, but the state attempts to fix or stabilise it to suit its own interests – that is, largely in terms of stability and efficiency (Parker, 1999a; Ravenscroft and Parker, 1999; Flyvbjerg, 1998; Jessop, 1990).

The basis and understanding that we have of citizenship can be as oppressive as it can be emancipatory. There have been extensive debates concerning the search for alternative political strategies and institutional arrangements in contemporary political science during the late 1980s and early 1990s, especially in terms of perceived inadequacies in the system of government and a lack of 'normative' engagement with political processes on the part of the vast majority of the public (see Healey, 1997). Power in the post-structuralist sense involves capillary action and, in Habermas's terms, the colonisation of the lifeworld (1987, 1988). Such colonisation is, however, undertaken not only by the state, but also by other powerful interests, typically via consumption practice and the media. On the other hand, it is also reflexively understood (albeit differentially) through a form of a 'mapping' process (see Jameson, 1991) and prompts alternative claims and actions beyond the accepted limits of the system.

Citizenship has been conceived of in the past as a role that we play out reciprocally through political elections or referenda in order for others to administer the mechanics of governance (Clegg, 1989). The doctrine of mandate in that system ensures that our collective interest will be cared for and the 'contract' renewed. Such a bargain also underlies Mauss's gift relationship (Mauss, 1990; Wilkinson, 1997; Parker and Ravenscroft, 1999; 2001) and hence its interesting application to the way that customary relations in the countryside operated, and both relate to social contractualisation. So, since the formation of the state in the early modern era there has always been a form of dialectic in the making and remaking of citizenship. Citizenship construction is a role that government undertakes and the state perpetuates, either as an acknowledged part of political policy or, more obliquely, as a result of policy that unwittingly helps construct or 'translate' citizenship:

> The republic does not simply leave the 'reproduction' of citizens to existing communities, but verifies whether the social formation enjoined by those communities allows for admission to citizenship. Where this is not the case or where the people lack the formative support of the community, the government interferes. The task of reproducing citizens is implied in every government action. Every government action can and may be examined in terms of its effect on (the reproduction of) citizenship, just as we now judge nearly all government action in terms of its effect on the financial deficit.
>
> (Van Gunsteren, 1994: 46)

Therefore citizenship is formed by action, and claims may secondarily rise from below (as well as beyond) the level of the nation-state. Van Gunsteren (1994, 1998) begins to consider the broadening of normative citizenship theory and touches on a third key point that has been explored by numerous commentators: that individual agency, local culture and, increasingly, new social movements develop their own interpretations of handed down rights and obligations. They develop resistances and produce their own codes which challenge dominant (national) constructions of good and deviant citizenship. Such appropriation and resistance is an echo at least of the traditional governance of the community based on locale (Giddens, 1979). In this sense at least, citizenship has been reflexive (see Lash and Urry, 1994; Urry, 2000). Further than this, more research should be carried out to trace how local culture translates into local and 'national' rights and responsibilities.

There have been sustained calls for constitutional change in the UK and attempts are ongoing to persuade government to amend the UK's constitutional arrangements. (It should be noted here that in legal terms, the people of the UK have not had the formal status of 'citizen', that being a title formally reserved for republicans. The UK has had neither a formal, collected or transparent constitution nor a bill of rights; see Klug, 1997.) It is apparent that purely domestic arrangements are being superseded. Recently the European Convention on Human Rights was finally incorporated into English law as the Human Rights Act 1998 (made effective in October 2000; see Grant, 2000; Johnston, 2000; Home Office, 2000; Parker, 2001), the full consequences of which are yet to be fully appreciated, but it is clear that more attention to particular rights will be paid to policies constructed across all public affairs. The Human Rights Act, it is claimed, is likely to ensure that recognition of 'rights' will grow (Pritchard, 2000). Rights can be even more clearly demonstrated as being, variously, constructed, exclusive, unusable and contingent. This legal point is contrasted with the cultural and political situation as regards citizen action and impact on politics in Chapter 8, where post-national and 'postmodern' citizenships are reflected upon in relation to the countryside.

Rather than citizens exercising periodic political agency once every five years and again during local elections (and then voting levels are very low, at around 30 per cent for local elections), groups and competing interests are engaged in an ongoing dialectic or process of argumentation using and practising politics daily in the form of 'diffuse politics'. Different groups, classes or factions will do this in different ways and may be observed differentially. It is also true that many people do not actively engage in the forms of political action that have normatively been considered part of citizen action described by Selman (1996; after Alinski, 1972) as 'civic sclerosis'.

For the most part, the reconstitution of citizenship is decided by conflict, contestation and developing the alternative forms of citizenship noted above. In relation to constitutional arrangements and citizenship, David Held (1989: 177) makes the point that formalised/state notions of citizenship are unlikely to be adequate to deal with the concerns of all citizens and that part of the role of the citizen should be to require accountability:

> If the state as a matter of routine, is neither 'separate' nor 'impartial' with respect to society, then it is clear that citizens will not be treated as 'free and equal'. If the 'public' and 'private' are interlocked in complex ways, then elections will always be insufficient as mechanisms to ensure the accountability of the forces actually involved in the 'governing' process.

This illustrates a tension between the development of 'citizenship' and the UK system of representative democracy and attempts to define citizenship in narrow and legalistic terms. They tend to serve a purpose of marginalising or 'othering' the actions, needs and views of a whole range of groups. The reflexive and processual citizenship model then argues that people are engaged, on a daily basis, in challenging the parameters of citizenship; perhaps subtly or unknowingly, they are altering the distributions of existing and nascent rights and responsibilities.

The establishment and curtailment of certain rights are essential parts of political projects, while others are residual historical fragments, or are subject to ongoing (local) cultural interpretation and challenge. The use of history is classically invoked by Habermas (1987) and underlined by Bender (1993: 275), who views the use of historical discourse as involving the 'mobilising [of] discourse differentially empowered through time'. Thus citizenship claims move up from below and are brokered, discussed and passed into (national) law. The nation-state attempts to consolidate and stabilise definitions offered and essentially homogenise, or assimilate, the population along the lines demanded by the definitions adopted, much in the same way that the enclosures were attempts to instigate uniformity and rationality in terms of land regulation, control and ownership structure. Flyvbjerg (1998) considers such rationality/rationalisation as an attempt to 'freeze' politics in line with the notions of stabilisation enveloped by Bruno Latour (e.g. 1987, 1994). Thus the state prefers, indeed requires, citizenship to be controlled and defined through the stable parameters of law. Otherwise it is claimed, pejoratively, that anarchy will ensue. It is suggested that the anarchic is reality and attempts to impose control and regulation on this state are attempts to regulate the 'chaos' of cultural form, identity, affinity, the imagination and shifting self-interest. To control the population would be a hugely complex undertaking that in practice is impossible. It is perhaps possible to maintain a critical dominance, however (see Jessop, 1990), where powerful or otherwise threatening groups are kept docile.

Historically, citizenship construction has been conceived of as a product of the social contract enjoined by the public and state. The historical context impinges on citizenship projects: some historical features or stabilisations are enabling while others are obstructive; much depends on the political doctrine adopted by the state. This flexibility cuts both ways and is a product of the lack of codified rights and a lack of a culture of certainty, transparency or accountability in the UK system (see Coote, 1998; Klug, 1997; Wright, 1994).

In Hobbesian terms the idea of citizenship represents the acquiescence or consents of the population (see Clarke, 1996; Sabine and Thorson, 1973). This simplistic assumption falls into a trap of assuming that citizenship comes from 'within' – from a self-referencing space. Challenges to such isomorphic construction therefore come from 'without' and should be treated with caution: alienation and lack of 'stake' are both a need and a threat from and towards the state. This point leads into discussions about the nature of community, identity and the durability of definitional boundaries (see, for example, Isin and Wood, 1999; Lash and Urry, 1994).

Liberal and communitarian conceptualisations of citizenship

British politics, especially since the First World War (see also Chapter 4), has been aimed at reconciling laissez-faire capitalism with progressive social democracy and building a state structure that mediated, predominantly, class-based concerns. It is one of the contentions put forward here that intermittent crises have prompted different policy responses from different UK governments; pertinently in terms of social investment and frequently in relation to land rights, management and use. Such crises allow for challenges to prevailing conditions to be made and for powerful interests to find ways of maintaining a hegemony based on manoeuvre: 'It may be ruled out that immediate economic crises of themselves produce fundamental historical events; they can simply create a terrain more favourable to the dissemination of certain modes of thought' (Gramsci, 1971: 184).

In terms of the focus of this work there are at least two forms, or categories, of crisis that impact on citizenship and the countryside – particularly land and governance. First, there is the general crisis that prompts calls for a total rethink about the relationship between the state and the individual – a political crisis. Second, there is the type of crisis that affects an aspect of the economy such as agriculture (as throughout the late 1990s) and similarly prompts calls for a renegotiation about the governance and use of land.

Citizenship is subject to manipulation by ideologically driven policy. This makes it important to set out two of the groupings of ideology in relation to citizenship, prior to detailing the political history of the projects of recent UK governments. First, *liberal* constructions of citizenship tend to emphasise the individualism of the citizen: 'the citizen as rational being, the calculating bearer of rights and privileges' (Van Gunsteren, 1994: 39). This form of liberal citizenship is theoretically calculated in order to result in the maximum benefit for the individual and relies to some extent on the individual meeting obligations and responsibilities rather than demanding rights and provisions (Dahrendorf, 1994).

The limits of this conception lie predominantly where the liberty of the individual compromises the liberty of another individual (Roche, 1992). The individualist notion of citizenship has recourse to philanthropy and the benevolence that has, for example, characterised much countryside access provision in the past, notionally to achieve 'social' objectives. The liberal version of citizenship leaves philanthropy and altruism as the remedies for disparities in conditions and access to opportunities – and not just in spatial terms – where the market fails to provide (Gyford, 1991; Dahrendorf, 1979). Under the liberal conception of citizenship the market mechanism is the predominant method of procurement of a citizen right and, influenced by the US citizenship construction, contractual relationships are emphasised (Fraser and Gordon, 1994). Under the market the 'citizen' has power by virtue of his/her role in the marketplace as a consumer (see Urry, 1995; Parker, 1999c).

A property right is a civil citizenship right under the present structure of citizenship. Under this structure, private property rights are defended against other citizen rights-claims. This renders exchange of rights through the political system problematic. Once installed, markets and 'values' become entrenched as the legitimate mode of regulation and transaction. Policies which operate according to market criteria have social ramifications for the community (of which the construction of citizenship is putatively there to protect); as Van Gunsteren points out, 'a community that is merely expedient is not a community' (1994: 41). Expediency is one of the keywords of the free-market discourse – the implication in Van Gunsteren's comment is that 'community' is a complex and often intangible construct exhibiting features such as altruism, compassion, helpfulness and identification with place. By inference here the composite phrase would be 'a citizenship that is merely expedient is not a citizenship'. The issue of the effects of markets on citizenship and community therefore concerns the expediency of the market and the 'value' of 'community' at local, national and even international levels of construction. The notion of citizenship and of community is one that is built on complex and sometimes fragile social, economic and historical foundations. The market rationalises these relationships. 'Value' in this sense is based on the willingness to pay for goods and services. Here it is contended that 'community' is theoretically constituted by individuals who are acting from degrees of self-interest (see Eder, 1993; Wright, 1994) and that the unproblematised notion of community itself should be treated with caution (see Gorman, 2000).

Second, the *communitarian* view expounded by Amitai Etzioni in the United States represents an attempt to rework the notion of the 'common good' (Mouffe, 1993; Etzioni, 1993) or the social/political model of citizenship. Gyford (1991) emphasises the role of community membership and sees the individual as being derived from that communitarian citizenship. The core of this model of citizenship combines, perhaps uneasily, the rights of the individual – labelled in terms of solidarity – with welfare rights (Dahrendorf, 1994). The community lives within a code in this instance, and while this code will necessarily be amended over time, it theoretically provides the framework for reproducing 'successful' or 'good' citizens. The model is dependent on the conscious creation (and re-creation) of a community. Unlike the liberal conception, this community conception is not reducible to individual agency.

Critics of the communitarian concept of citizenship argue that the basis of 'majoritarianism', upon which the US version rests, does not appear to allow for the empowerment of minority groups. The concept still seems to exhibit liberal or 'modernist' tendencies towards assimilation. If a concept of citizenship could be negotiated that integrated diverse opinions and views, it would represent a fundamental reconstructive step. Amending and re-working the common code to accept other rights claims would result in a society tentatively labelled 'neo-republican' by Van Gunsteren (1994) or one that could be called 'liberal socialist', where society attempts to construct a pluralist democracy (Mouffe, 1993). Etzioni (1993) claims that there are particular strategies that are part of the communitarian agenda – for example, the devolution of power to local areas, and the further democratisation of social relations (see also Hutton, 1997; Ackerman and Alstott, 1999). Many of these ideas have been incorporated into New Labour policy in the UK, while the Major government's 'back to basics' idea appeared prima facie to adopt a communitarian 'code' – surprisingly similar in tone to the approach of the succeeding Blair administration. It is argued that the construction of communitarian citizenship is rendered problematic by the diverse nature of a culturally fragmented plural, or postmodern, society in similar fashion to that of liberal constructions (see Heater, 1990).

Current policy efforts in the countryside are linked yet constrained by many external influences – for example, the reform of the CAP – and, ubiquitously, the influence of the land and agricultural lobby remains strong over central government despite shifts in the economic significance of agriculture and related enterprises. The common labelling of recent Conservative administrations as liberal comes about in recognition of the emphasis placed on a 'free' market economy and a minimal state within government policy since 1979. Cox (1984) emphasises that policies to return land and property markets to a free market have failed in the past, as have efforts to nationalise land. The power of constraint, or Denman's (1978) 'positive power' of property ownership held by landowners, constrains the power of policy initiation held by the state, which in turn renders radical progression along routes towards free markets or land nationalisation unlikely. In the case of contemporary policy initiatives for agricultural land use there is an underlying shift to address land use to market demands. These shifts are felt particularly in terms of public access provision to the countryside. Access is sensitive in this way because of its fragility in legal *de jure* and civil terms. In terms of economic arguments, countryside access is recognised as a means of income generation for the countryside through tourism and associated spending on goods and services located in the rural.

Citizenship may be a status and might also imply forms of socio-political participation on the part of the individual, but it is also a *resource*. In terms of this notion of citizenship as resource, Foucault would regard both the identity and the status of citizenship as representing a reservoir of power and of governmentality. It represents a definition that is under constant challenge from different claim groups and a tool for mobilisation of an interest position. The translation of claims into rights, obligations (and the rejection of claims), undertaken by the state, needs to remain relatively stable for the system of governance to operate as it has done – that is, largely top-down prescription and centralised analysis of problem/solution. Simple engagement with a single interest group can also, however, lead to a corruption of the idea of reflexive government as one powerful lobby seeks to preserve its channel of influence (Grant, 1995). It is only where definitions are shared that co-operation tends to develop (as with agriculture in the post-war period; see Winter, 1996) and perhaps in terms of housebuilding in the 1980s (Murdoch and Marsden, 1994). Given this line of argument, the use, construction and definition of citizenship enjoined by UK government over the past two decades are examined in the following section with reference to the influence of organic citizenship influenced from beyond the national.

Political projects and citizenship rhetoric

Citizenship has been the preserve of political theorists, but increasingly the notion has been rediscovered as a powerful tool for politicians within and outside formal politics. This is perhaps because the term appears to hold little or no negative connotation (Marston and Staeheli, 1994; Cooper, 1993). For the most part it is viewed as a positive symbol of status. It is not surprising, then, that UK politics and leading politicians have sought to utilise citizenship rhetoric as part of their particular political projects in the 1980s and 1990s.

There has been a recent history of extensive and explicit political exhortation towards 'citizenship' as well as actual construction or baseline engineering of citizenship that is part and parcel of political change – as encapsulated by Van Gunsteren (1994: 46). In this sense, citizenship envelopes will shift towards one political pole or another without presenting a marked movement towards one conception of citizenship or another. Gramsci (1971) identifies this as part of the 'conjunctural': the everyday of political action. It is highlighted here, and expanded upon later, that law and order and the dilemma of balancing freedom with control is a constant theme in the citizenship discourse and one which government is continually reworking and, from a variety of sources, being pressured to rework.

Express use of citizenship rhetoric can largely be viewed in terms of the desire to regulate freedom: requiring the public to do such and such and not to do other things, in line with existing legal frames, policy manifestos and the ongoing balancing of domestic budgets. The second aspect to this is the paying of lip-service to the idea of enhanced political participation at the national and local level (cf. Gyford, 1991; Burns *et al.*, 1994). This is essentially a clarion call made in order to persuade the people to fit with the system rather than the reverse, or a composite, perhaps negotiable, system. Neither of the main political parties in the UK has seriously considered systems such as proportional representation. Even though the Labour government set up the Plant Commission to report on such matters, its recommendations have been discreetly shelved (see Institute for Citizenship Studies, 1999).

Recent Labour attempts at government devolution to Scotland, Wales and Northern Ireland have been criticised because of central attempts to control and restrict important aspects of their operation, including the selection of leaders – a controversy in the first London mayoral election. Various administrations of differing political hues implicitly order different types of right and obligation, emphasising and prioritising those that rest most comfortably with their political viewpoint. This is done at the expense of unadopted rights-claims, and often disadvantages minorities and minority-held viewpoints. The role of the state in determining legitimate rights and responsibilities (as opposed to legally enshrined rights and obligations) is a crucial one. However, particular rights and their classification are justified and legitimated only under particular conceptions of citizenship or outcomes associated with allied political projects. Hutton (1997) notes that the Majorite response to the preceding Thatcher years was to adopt a no-change position in many areas of rural policy, particularly as agriculture went through a short-lived period of relative affluence in the mid-1990s.

There are a variety of factors that influence decisions regarding the citizenship 'envelope'. These include (generally and specifically in relation to the subject matter developed here) economic impact, moral argument, crime levels, levels of environmental degradation and, crucially, the effect on existing rights or obligations. Bauman, reflecting on Sennett's research into US towns, claims that their stance, looking out into a world of uncertainty, risk and destabilisation, was to retreat to a 'bunker mentality' where 'the suspicion against others, the resentment of strangers, and the demands to separate and banish them as well as the hysterical, paranoiac concern with "law and order" climb to their highest pitch in the . . . most homogenous communities' (1998: 47).

There is a danger, as traditional rural interests respond to change and attempt to acknowledge and build a more diverse countryside, that government will listen only to embedded or powerful voices (Fairlie, 2000). Giddens (1998) echoes this when he claims that to embrace and engage with the effects of globalisation is the only way forward. To ignore it, he warns, is not an answer that can be contemplated unless the world is to split, Orwellian-like, into warring blocs; dividing the UK into Rural and Urban has similar effects on all parties. A careful analysis of what is required by a mobile and diverse society is crucial for the UK as a whole, for its constituent nations and for localities within them.

Government and rights

After the exit of Margaret Thatcher as prime minister in 1990 the Conservative Party in the UK suffered destabilising factional differences reflecting the existence of very different camps of conservative Tories and neo-liberals. The two had coexisted, albeit uneasily, during the 1980s under her 'iron' leadership. This unity enabled the New Right agenda of the 1980s to proceed with a rallying cry of 'free market and strong state'. In relation to the latter aspect of this there has been one point of continuity through the Thatcher and Major years and now into the Blair project (with its discursive turn to community that was begun in part by John Major): one continuing feature of government policy has been to react to rising and perceived rises of crime, and essentially the expression or performance of alternative rights-claims. This threat has been countered by enacting rafts of public order legislation that have largely been aimed at 'others' (in some sense 'aliens') who have been cast in the role of threatening 'non-stakeholders' (or perhaps those who become *too* 'active') and at the protection of private property. As Thompson states in the opening paragraph of *Whigs and Hunters* (1973: 21), 'there are more ways than one of defending property'. The Terrorism Act 2000 and the Criminal Justice and Public Order Act 1994 are recent examples (see Home Office, 2000; see also Fyfe, 1995; Bucke and James, 1998; Parker, 1999a). Such measures are latter-day reflections of the eighteenth-century Black Acts described by Thompson (see Chapter 4).

The enforcement of law and order policy, which in recent times has been used selectively by government and the police (see Bucke and James, 1998), illustrates how rights become implicitly gradated in accordance with the dominant ideology, coupled with the dominant power in the legislature. Dahrendorf (1994), arguing from a social democratic viewpoint, thinks that rights can be conceptualised in terms of their relative necessity or 'embeddedness'. He states that there is a 'hard core' of rights that are 'fundamental and indispensable' (*ibid.*: 13). This may be so in order for society, as we presently conceptualise it, to function and for key interests to remain stable. In theoretical terms, however, no such rights are prerequisite or immutable. Held (1989) and Giddens, as mentioned in the previous chapter, forcefully argue that rights have been won through argumentation and require continual defence on the part, or on behalf, of those who benefit from them. This does not mean that they should not be contested and debated. Doing so is an important part of democratic renewal and of healthy dissent. The problem can be seen as being exacerbated by rights and responsibilities distributions that favour powerful groups and a system of citizenship formulation and regulation that favours dominant groups. Such interests are ready and able to protect their existing rights in order to resist harmful change.

Rights can more usefully be viewed as contingent, but one of the functions of certain rights – previously termed 'jostle rights' (Parker, 1996, 1999c) – is to provide legitimate spaces for resistance and political debate and to preserve the opportunity for quasi-formalised political exchange. Such rights are increasingly mediated, thus avoiding contests over physical space by using new technologies, such as the Internet and other media forms (see Mobbs, 2000). In some instances protest is organised and amplified using the media, but direct confrontation is also, seemingly, on the increase as more groups challenge dominant constructions and locate each other via the new technologies. It is ever more the case that direct action provides media spectacle for rights-claims to be promoted. Such technological advances enable more organised challenges to extant (formal) citizenship. Numerous recent examples that have taken place in and over rural issues include the protests in East Anglia against genetically modified crops (*Guardian*, 20 April 2000), and the 'guerrilla gardening' protest at the London Mayday 'carnival' (*Guardian*, 2 May 2000). Chapter 5 assesses attempts by citizens to influence government and local government through organised forums or processual arrangements, and protest and direct action as forms of 'active' or 'engaged' citizenship are explored in Chapter 7.

In terms of political history, Macintyre (1999) notes, in a recent review of 'stakeholding', how Lloyd George began in some way to develop policy for a more equal society with his people's budget of 1909 (Douglas, 1976; Cherry and Rogers, 1996). In the 1920s Chamberlain recognised the political and social benefits of a more inclusive approach. For example, his maxim 'every spadeful of manure dug in, every fruit tree planted' had the effect of converting a potential revolutionary into a citizen is a clear indication of this (quoted in Macintyre, 1999: 127). This exhibits shades of the seventeenth-century Diggers' rhetoric, as explored in Chapter 5. In the 1930s the Conservative politician Harold Macmillan, later to be prime minister, had espoused a 'Middle Way' that appears now to be very similar to the 'Third Way' that is currently under construction by New Labour. Even though the economic and social context was rather different, the notion of 'One Nationism' is firmly back on the political map – of England at least. The building of citizenship under the Macmillan model was to be constructed around twin pillars of education and capital accumulation: to ensure a fit and educated workforce and the security of the interests of capital. This is clear ancestry from which the current UK political approach stems.

In introductory fashion, each of the last three government approaches to the relationship between the state and the individual, if not explicitly citizenship, over the past twenty years, are set out. This includes policy that has explicitly used citizenship rhetoric as well as the more general rhetoric and underlying political philosophies. It is also of particular interest to review how socio-economic change and associated theorisations of citizenship have been reflected in such projects. Of the three phases, the Major (1990–1997) and Blair years (1997–) are given prominence here. They have both employed citizenship rhetoric overtly and as central parts of their political–social projects. The New Right project of Margaret Thatcher is a key antecedent and counterpoint to both Blair and Major in terms of their constructions of citizenship. This is especially the case because Thatcher attempted to redefine, in narrower terms, appropriate relationships between state and citizen, denying, if laconically, the existence of anything as *collectif* as society.

Margaret Thatcher, 1979–1990 – 'there is no such thing as society' and the New Right

The prime concern of Conservative governments prior to the Thatcher administration had been to protect economic rights, to defend the status quo regarding key rights distributions (such as property rights) and to minimise change. There had also been a desire on the part of the Conservatives to favour a benevolent yet selective corporatist approach. Where new claims or circumstances challenged embedded interests, the established economic interests would generally be favoured (Sabine and Thorson, 1973; Hutton, 1997).

Breaking with the post-war political 'consensus' in British politics (Cox, 1984; Thornley, 1993; Giddens, 1995; Macintyre, 1999), the Thatcher government aimed radically to alter the trajectory of UK politics away from welfarism and away from 'community politics' (Denman, 1978). Symbolically, one of Margaret Thatcher's most famous pronouncements, made early in her period of office as prime minister, was to deny the existence of society. This, rather contrarily, anticipated the constructed rather than non-existent nature of society. She simultaneously denied other social constructions and selectively recognised other 'realities' and imperatives. That was very much her conceptualisation of politics – at the very least, such selectivity was relatively transparent. The government in this period was certainly schizophrenic in many respects, and in terms of citizenship Fyfe (1995) makes a strong case in arguing that the Thatcher government employed a mix of conservatism, advocated by 'Tories', with hefty doses of the rather different political creed of neo-liberalism, advocated by the New Right (for a more refined explanation, see Rhodes, 1988; Gamble, 1988). The citizen existed as an individual within a society that was seen as largely homogeneous and based on 'tradition, consensus and the rule of law' (Smith, 1989: 146). Numerous commentators in the 1980s and early 1990s labelled this particular dogma as 'New Right' (King, 1987). The New Right were concerned much more with economic structure than with developing citizenship and associated political restructuring. This was especially so since, by the 1983 election, Thatcher had won a record majority and seemed politically unassailable. This also meant that she managed to silence the one-nation Conservatives in her own party and maintained the backing of the spectrum of political ideology housed in the Conservative Party. In the ensuing years the Conservative Party reworked almost every regulative system, including the relationship between workers and employers, and the role (if not the framework) of the planning system (Thornley, 1993), and carried out an extensive privatisation programme.

Another outcome of this political fusion, in terms of citizenship, was to place the emphasis squarely on the individual to be a 'responsible citizen' – in essence, to be responsible for oneself first (Macintyre, 1999). Implicitly this self-interest should rationally be in the interest of others – a directly parallel and reductionist argument used to underpin the 'trickle-down' politics of Ronald Reagan's 1980s United States. This also presumes a particular moral code and ignores or at least undermines alternative and minority interests or cultural expression – in one breath denying society, in the next, implicitly relying on social networks and notions of 'social responsibility' to underpin processes of economic restructuring. Where these were not sufficiently in evidence, then the legitimate role of the state would be to enforce strict law and order policies (for example, by the 1986 Public Order Act; see Waddington, 1994). Together they represented the maxim, as mentioned above, of 'free market and strong state' (Gamble, 1988). It was in this context that Thatcher welcomed the notion of 'active' citizenship – a very limited envelope indeed.

The Conservative Party challenged the basis of land-use planning, presuming that all development proposals were permissible until proven otherwise (Ambrose, 1986; Thornley, 1993). This prompted a rise in planning permissions, notably on greenfield sites (Murdoch and Marsden, 1994; Healey *et al.*, 1988). Complaints from rural dwellers gave rise to the now famous 'not in my back yard' accusation made famous through the NIMBYist tag (see

Burningham, 2000) and the Environment Secretary who promoted a looser stance towards development, Nicholas Ridley, who was discovered objecting in NIMBYist fashion to development near his own home.

The restructuring of agriculture and its regulation (Winter, 1996) had always been a notable exception to the scope of the planning system, and one that is of importance here (see Cherry and Rogers, 1996). In some senses the 'free' regulation of agriculture was already made extremely difficult by membership of and participation in the European Common Agricultural Policy. The task is, arguably, beyond the control of any single state, given the advent of the global economy (see, for example, Ray, 1999). To regulate land use and change all aspects of agricultural and related use would be a huge undertaking and one that would have run contrary to the Thatcher project, but even the dominant New Right faction of the Conservative Party had to be careful not to destabilise the 'Tory shires' too much. Even with a large majority, Thatcher had no desire to antagonise traditional conservative support in the shires by observably altering the status quo in rural areas. The Conservative Party drew a mainstay of political support from the traditional economic and cultural base of the rural. This had always stemmed from agriculture and associated industry, and the Conservative domination of areas beyond the major towns and cities provided a strong power base (Johnston *et al.*, 2001; Winter, 1996; Rhodes, 1988). The shadow of the Scott Report still held enough power over the Thatcher government for its main implications to prevail, especially in the early Thatcher years. It was, however, becoming apparent that serious flaws existed in the agricultural system (see Winter, 1996; Thirsk, 1999).

Significantly, little was done about public participation in public policy-making despite increasing levels of disruption, dissent and dissatisfaction with the government about many aspects of its policies. This, noticeably, included large minorities inside and outside of the Conservative Party who were unhappy about how the government – particularly in terms of planning policies – was not dealing with rural and environmental issues. It was only late on in the Thatcher era that the Conservatives attempted to tap into the groundswell of environmental concern. The Green agenda had steadily been rising around Europe, notably in (West) Germany. At the European elections of 1989 the green vote was significant. For example in Hampshire the Green Party took 20 per cent of the vote (Thornley, 1993). There was already a strong 'wet and green' faction within the Conservative Party who had consistently lobbied for the government to adopt a more restrictive, ostensibly 'light-green' approach to certain environmental impacts, notably greenfield development.

One of the main policy documents relating to rural issues (there was no overarching rural policy during this period) was the White Paper *Our Common Inheritance*, published in 1990. This marked an attempt to persuade the electorate of the green credentials of the Thatcher administration. In terms of our focus, the document outlined how the government would place environmental considerations higher up the political agenda. This proposed change of policy of course held a whole range of implications for the countryside and the way that land in particular was to be used in rural areas. This led to a softer line on the environment, whereby environmental impacts would be accorded more weight in decision-making, an approach adopted by John Major when he took over the reins of power in late 1990. The new approach is reflected in the changes made in the planning legislation in the 1990 and 1991 Planning Acts (see Thornley, 1993; Cullingworth, 1999). Development plans were reinstated as the prime consideration in assessing planning applications – at least partly acknowledging the importance of a form of local control over land use (Thornley, 1993). *Our Common Inheritance* made reference to the importance of protecting certain environmental assets and made explicit mention of the water meadows surrounding Salisbury Cathedral. Only a few years later it was revealed that plans had been in preparation to build a new bypass around Salisbury that would pass through the very meadows referred to in the White Paper. By 1996 the road scheme was set to become the scene for the biggest and most bitter protest – even higher-profile than the Newbury protests (see Parker, 1999c; Merrick, 1996; Brown, 1996a, b). The protest was averted when the scheme was shelved almost immediately after the election of New Labour in May 1997 (see Chapter 7).

John Major, 1990–1997 – benevolence and the 'active' citizen

In 1990 Margaret Thatcher's own party, who feared that she had become unelectable in the face of a rejuvenated Labour Party under Neil Kinnock, toppled her. Her successor was her own nominee: John Major. Initially the momentum of the New Right 'self-help' ethic continued from the Thatcher years, and little of the Thatcher trajectory appeared to change. Will Hutton (1997: 2) goes as far as to say that Majorism was the 'bastard child' of Thatcher. At first Major was still in charge of a party that espoused much of the dogma formulated by key political architects of 1980s politics (such as Keith Joseph and, of course, Thatcher herself). Major struggled to reconcile the factions within his own party, referring famously to several of his own cabinet members as 'bastards' because of their conflicting and scantily concealed New Right credentials. However, as indicated, a less pro-development line in terms of land and planning was pursued with the advent of the 1990 and 1991 Planning Acts, which re-established the primacy of the development plan. This of course tipped more power back towards local authorities (see Cullingworth and Nadin, 1997).

Major settled into his position, narrowly winning the election in 1992. This enabled him to further develop his own approach and push his own policies forward – the Citizen's Charter had already been promulgated (see Cooper, 1993). In particular, it became clear that he wanted to reinvent citizenship and develop policy that explicitly played on the cultural capital or symbolic meaning that the notion of 'citizenship' held. The New Right approach was leavened by Majorism. The notion that citizens should actively assist at the level of community (rather than the individual) was proclaimed. Major utilised old conservative values of responsibility, charity and benevolence to promote his notion of 'active' citizenship – an approach that played heavily on the ethic of voluntarism (Wright, 1994; Kearns, 1995). An explicit call for rural people to be active citizens is exemplified in the rural White Paper of 1995, 'encouraging active communities which take the initiative to solve their problems themselves' (DoE/MAFF, 1995: 10).

One of the examples used in Chapter 6, the Parish Paths Partnership scheme (Parker, 1999b), was explicitly mentioned and exemplified as an example of the preferred activity of Majorite construction of citizenship. This version of active citizenship also promoted the idea of responsibility and of obligation. Importantly, the connection was made, albeit in limited fashion, between legal citizenship and moral/cultural citizenship, whereby a range of motives were expected, not necessarily based on self-interest. The Major government was expressly exhorting and mobilising the public to take part in public affairs – as part of a wider social contract – and was using citizenship rhetoric to do so.

Major's notion of active citizenship was largely about people helping others on the basis of altruism, and encouraging the individual to complain when certain public service standards were not met (Kearns, 1995). Such a limited view of citizenship meant that citizens were not to be encouraged to engage actively at the level of politics or to seek to alter structures and policies, but to act as monitors for existing policies and services. Tony Wright argues that the Major government was not at all interested in reinvigorating a critical or challenging form of citizenship where politics is 'learned and constantly practised, in arguments, activities and institutions' (1994: 55). So the role was constructed as being one of ensuring that existing entitlements were delivered properly. This was to be the main point, rather than to claim new entitlements or to be actively involved with governance.

In terms of policy vehicles, Major put in train several monuments to his rather shallow conceptualisation of active citizenship (Clarke, 1996). For example, his so-called 'big idea' was to introduce the Citizen's Charter in order that governmental and quasi-governmental agencies particularly would operate to set standards and pledge certain responsibilities to the public. Examples involved promising refunds if rail services did not meet service standards – the active part being that the individual would have to complain to the relevant authority to receive compensation. Such measures have continued in some sense under the Blair government with the push towards Best Value and benchmarking in public services (DETR, 1999; Ravenscroft, 1998). These are not intrinsically bad measures, recognising at least that institutions do need to inform people of their rights, but they hardly constitute a deeper formulation of active political citizenship.

Major explicitly encouraged citizenship education through a cross-curricular theme of citizenship, recognising that education was an important part of 'becoming a citizen'. Citizenship education has been maintained and encouraged by the Blair administration, with Citizenship becoming a compulsory subject in English and Welsh schools from 2002. It is a thread that, Bourdieu would argue (and as was noted in Chapter 2), has existed and played a crucial part in citizen formation from the beginning of compulsory state schooling (see Bourdieu, 1977; Institute for Citizenship Studies, 1999). After Baudrillard (1981, 1993), such 'diffuse education' is increasingly mediated and learnt through the sign economy. In this sense, citizenship becomes as much cultural as it is political or economic. This theme is returned to later where consumption and practice are discussed.

Major's project was politically and philosophically narrow. Kearns (1995) notes that the government did not wish to hear a politically vocal citizenry, rather that children should have their societal, and most particularly their political obligations emphasised to them while at school. The cross-curricular theme is an explicit and ideological recognition that through education people may be tutored to behave and consume in particular ways (Donzelot, 1980; Bourdieu and Passeron, 1977). This project is also reflected in the agenda of the Institute for Citizenship Studies, which has concentrated on engaging with state-defined structures of participation and gentle exhortation to business to exercise corporate citizenship (Institute for Citizenship Studies, 1999; Institute for Citizenship, 1992; McIntosh *et al.*, 1998).

In many senses the project of Thatcherism did continue to drift through the Major years as the legacy of such a strong ideologically driven period of office. The political philosophy of both administrations allowed for an unimaginative conception of citizenship. Ravenscroft (1993) argued that Major's signature was the disingenuousness with which his 'active' citizen would allegedly operate in a 'classless society', omitting to concede that such a classlessness could arise only as a consequence of 'constructed omission . . . where choice [is] replaced by means' (p. 33). As has been signalled already, statist forms and definitions of citizenship can be drawn arbitrarily – much as plans and planners (other arch-modernists) draw lines across space, attempting to bound possibilities and to separate and intervene between, and for, both the state and capital. In this context, capital is understood widely (after Bourdieu) as involving the 'capacity to act' or, more prosaically, to 'get something done' (cf. Giddens, 1998; see also Falk and Kilpatrick, 2000).

The social contract was quite clearly to be, first and foremost, economically productive, and second, charitable. In terms of rural affairs the Major years saw the first rural White Paper since the 1942 Scott Report, *Land Utilisation in Rural Areas* (MHLG, 1942). In 1995 the White Paper *Rural England* was published and a partner document for Scotland was also produced (DoE/MAFF, 1995; Lowe, 1996; Hodge, 1996). Both were roundly criticised by a broad range of rural interests who had been waiting hungrily for central government to supply prescriptions for what appeared to be a whole range of rural ills: declining rural services, declining farm incomes, greenfield development, breakdown of community – all were high on the agendas of interest groups, academics and significant sections of the public. It seemed that all the issues required more resources, different priorities and certainly a radical rethink about how rural affairs were conducted. Lowe (1996) and Murdoch (1997a) both contend that no new money was made available for rural policy and that 'extreme public expenditure constraints . . . bedevilled the whole rural policy review process' (Murdoch, 1997a: 116). This claim could be levelled once again at the 2000 rural White Paper, but as we shall see, among a plethora of policies it does attempt to enliven local-level politics and engagement through a policy of invigorating parish councils (DETR, 2000; Anfield, 2001).

Tony Blair, 1997– 2001 – New Labour and 'engaged' citizenship

Traditionally the Labour Party in the UK has championed the cause of progressive leftist politics. One of the central planks of the socialist programmes since the first Labour government of 1924 had been to establish more extensive civil, political and social rights. The project was largely an attempt to remedy perceived inequalities and imbalances in terms of social and economic opportunities and safeguards (Macintyre, 1999; Briggs, 1961). The focus of attention was on the extension or protection of (collective) rights. Rights were therefore seen very much as an important part of ensuring certain equalities, both in terms of protective rights and in terms of entitlements – particularly as regards education and health. As discussed in Chapter 4, rights that enabled people to play a part in determining the way that they would be governed were also important, with changes in government structure, voting rights, and arrangements for formal public participation being extended during the nineteenth century. This was brought about largely through pressure from the Labour movement and organised protests from a range of interest groups and individuals (see Chapter 4; Mingay, 1989; Sabine and Thorson, 1973; Douglas, 1976).

Following eighteen years of unbroken Conservative rule the Labour Party came into office in May 1997 proclaiming its own big ideas. Much had changed in terms of global economics and social attitudes since the last UK Labour government departed office in 1979. Culturally and economically, Britain and the nation's place in the global context had altered radically. The Labour Party went through difficult times in the 1980s attempting to adopt a political model that would be appropriate and acceptable both to its members and to the electorate at large. Tony Blair's 'New Labour' vision for the UK has been associated with communitarian theory (see Philips, 1993; Etzioni, 1993) and one of the key buzzwords to emerge was that of the 'stakeholder' and of 'stakeholding' whereby each person, each stakeholder, has a social, economic and political obligation to the state. Reciprocally the state has a responsibility toward the citizen, essentially for rights to be matched bidirectionally with responsibilities. This approach calls for an emphasis on community and for claims for individual rights to be assessed against those of the community. It also appropriated an explicit concern with 'citizenship' (see Plant, 1994).

The New Labour approach towards citizenship was never more apparent than in Tony Blair's 1997 annual Labour Party conference speech, made just months after a landslide election victory, where he spoke of individual citizens matching their rights with responsibilities: 'a decent society is not based on rights. It is based on duty . . . the duty to show respect and tolerance to others' (Blair, 1997: 14). This neatly illustrates the New Labour emphasis on duties, if not explicitly upon action/activity, rather than prior Labour administrations' focus on rights provisions. For the first time, Labour was explicitly acknowledging the conditionality and reciprocality of citizenship.

In hindsight, the notion of citizenship as constructed by John Major was not too far removed from this. It seems that the Majorite active citizen is still encouraged under New Labour, and more jargon for citizenship has partnered the stakeholder rhetoric: the notion of the 'engaged' citizen has been put forward to incorporate the criticisms that the Majorite form was not a political construction. The Blair view centres on the idea that:

> the notion of active communities and engaged citizenship is crucial . . .
> the rights we enjoy reflect the duties we owe. The desire for rights cannot be separated from our responsibilities to each other, to society and to the future. Today we must confront how we can work together as citizens to create active communities.
>
> (National Policy Forum, 1999: 68)

Hence, underpinning the Blair project is a notion of national morality as the intermediary of rights and responsibilities. As the Blair project has unfolded, this type of rhetoric has intensified. Citizenship and 'stakeholding' have become central to the project of 'modernisation' envisaged by Blair. Part of this modernisation is, in Giddens's (1998) terms, about renewing social democracy. For Blair, this mobilising discourse has become one of matching rights with responsibilities, much as Major had done. It appears that there is little difference in substance between the philosophy of Blair and Major, although one important difference has been in the style and delivery of the Blair project.

The label 'stakeholder' requires further unpacking because it has important connotations, as mentioned in Chapter 2 (see Hutton, 1996, 1997; Ackerman and Alstott, 1999). Anthony Giddens, as one of the architects informing the building of Blair's 'Third Way' politics, in his book *The Third Way* effectively produced the user's guide to the New Labour approach: 'Third Way politics looks for a new relationship between the individual and the community and a redefinition of rights and obligations. One might suggest as a prime motto for the new politics, *no rights without responsibilities'* (Giddens, 1998: 65). Such a movement may represent a shift back towards the consensus politics of the post-war years, but Macintyre (1999) adds a cautionary note: that the rhetoric obscures an attempt to consolidate elements of the Thatcher programme in terms of the continuing retreat from the state provision of welfare. One conclusion that can be drawn is that citizens are increasingly encouraged to make 'social' contracts with a range of organisations and structures, all with varying motives, scales and locations – for example, through private healthcare as a means of minimising risk, or the use of education primarily as a cultural token, one to be deployed in order to find work. Postmodern citizenship is about making a social contract not only with the state but with other sociations at varying scales. This leads citizens to find their own ways to register politically and economically.

Different notions of the 'stake' can be assessed and be viewed as appropriations – all within the embrace of the social contract. As discussed, the historical antecedence of the 'stake' is long established – especially and pertinently in relation to land and rural space. As long ago as the 1640s, when the English Civil War had brought issues of proper governance out for open debate, claims for each person to have a 'stake' were put forward as part of the push for a literal commonwealth. This was 'radicalism born of growing disillusionment and despair at the inability of the national leadership to give the poor and property-less a stake in the new-born republic' (Boulton, 1999: 43). Most recently, and perhaps reflecting the dominance of capitalism, Ackerman and Alstott (1999) suggest that in the spirit of true stakeholding, all US citizens should be credited with the sum of $80,000 – which would be paid back over time to the state. In similar fashion, in the UK at least one group has been lobbying government to guarantee a 'citizens' income' whereby all nationals would receive a basic payment as a form of stake (Citizens' Income Trust, 2000). The implications of such suggestions are beyond the scope of this volume, but it is interesting to note how concepts such as the stake are being variously applied.

In terms of domestic politics, national government in the UK, since the mid-1980s especially, has sought to re-engineer political responsibility as one response to intensified political scrutiny by the media and the public by setting up a range of quangos. It has done so partly to ensure that central government priorities are achieved by circumventing local politics, and partly to put the accountability for the outcomes of the policies operated through those bodies at arm's length. A form of this type of dynamic might be seen through the Highways Agency's role in carrying out the roads programme of the previous Conservative governments. Associated protests (to protect the 'rural environment') of the early to mid-1990s struck not at the government, or indeed the agency, but at the civil engineering firms bidding for the road-building work, as is mentioned further in Chapter 7.

Indeed, risking further loss of control, but endorsing local participation and invoking notions of 'stakeholding' (DETR, 1998a, c; Giddens, 1998;

Ackerman and Alstott, 1999; Hutton, 1997), New Labour has moved to devolve powers to Scots and Welsh parliaments and to reinstate the Assembly for Northern Ireland. Devolution has, however, met with some controversy as the centralising instincts of national politicians wrestle with the idea and the reality of allowing regions and localities to choose for themselves (and raise their own funds), at least in some aspects of policy. The controversy has led to accusations of 'control-freakery' on the part of the Blair government, just as there was strong criticism of Thatcher and her centralising tendencies in the 1980s (Stoker, 1988; Ambrose, 1986). Hence there has been a continuing tension between, on the one hand, rhetoric of citizen 'empowerment' and, on the other, control remaining with central government and a burgeoning range of quangos.

Attempts to ensure that candidates form leadership positions that are sympathetic to the Blairite approach have met with opposition. In Wales this opposition was successful in forcing the resignation of the Blair candidate Alun Michael from his office as leader of the Welsh Assembly. Similarly, problems arose when Ken Livingstone, a Labour critic of the Blair programme, was discouraged at every opportunity from standing as the Labour candidate for the newly created mayor of London post. He announced his candidacy as an independent after being narrowly defeated by a top-heavy selection system that was designed to favour Blair-sponsored candidates, and eventually won the mayorship in 2000. Changes of this sort in national/regional governance are in their infancy, but it is likely that relations between such assemblies, Westminster and Whitehall will be tested in the coming years.

Since the 1997 election the government has created somewhat of a paper-storm with wide-scale consultations and discussion documents about many aspects of the countryside (cf. DETR, 1998a, 1999). There has generally been a move to reorganise and instigate a culture of joined-up thinking (see Ward, 1999). The conceptualisation of citizenship has seen some interesting outcomes for the countryside. One policy included in the rural White Paper for England (DETR, 2000a) was to ensure minimum service standards for rural areas to redress the poorer provision of entitlement in the country in relation to urban areas; this is an attempt to redress an obvious failing in the universality of citizenship across the country. In terms of governance, a move to build on a Majorite policy of empowering and placing responsibility onto parish-level administration was announced under the label of Quality Parishes (see Anfield, 2001). In the partner urban White Paper (DETR, 2000b) Community Strategies were announced as the only new channel for participation in urban/rural policy, also reflecting a rather feeble attempt at 'empowerment'.

There has been an emphasis on partnership and 'social efficiency' through attempts to integrate and co-ordinate rural policy and how things are operationalised in the countryside, under the labels 'joined-up thinking' and 'joined-up policy'. In the consultation documents that have been produced on

rural issues (DETR, 1998a, 1999), some clues as to how the stakeholder economy and society will be mediated in the rural can be discerned. In the consultation document issued by the government (DETR, 1999) there appears to be
little of substance that is different from the 1995 Paper except the tendency
that is developing to 'join things up'. In the widely read Performance and
Innovation Unit report *Rural Economies* (PIU, 1999), a number of pointers
towards Labour priorities may be detected. These include the notion of
encouraging enterprise, of 'joining up' services, better access and accessibility
both in terms of land and services, and similarly a stronger linkage between
urban and rural.

Again the policies outlined in *Our Countryside: The Future* (DETR,
2000a), while implying a greater role and level of engagement for the public
qua citizen, on its own (and when compared to the urban White Paper
issued just previously (see DETR, 2000b)) it does not seem to be capable
of implementing the necessary interventions, regulation and freedoms that
a diverse, sustainable and working countryside needs. Attempts to devolve
responsibilities – and maybe some powers – to the hands of local people,
perhaps at the parish level, and to encourage voluntarism are continuations
of processes put in train by John Major (see DETR, 1998a, c; Countryside
Agency, 1999a). It is too soon to see the extent to which Blairite policies
(and the economic approach of the chancellor, Gordon Brown) will impact
on the citizen and on the rural. It is rather early to assess the Blair agenda
for the countryside, but it seems clear that radical state-sponsored change
will not be forthcoming. The rural White Paper published in November
2000 did not address many of the seemingly deep-rooted or structural
economic issues, for example, that face British, European and global agriculture (see Murdoch, 1999) and that affect many aspects of rural and urban
life. Other flows are seemingly impossible to arrest, and the demand for
housing in rural areas is still further transforming many areas of the countryside. New-build is limited, and inmigrants tend not to use local services
and amenities, thereby undermining them.

Attitudes of governments towards the shape, extent, style and legitimacy
of such rights are very important. Such attitudes can be examined through
a variety of avenues in a form of grand discourse analysis. In UK rural studies,
such attempts to review government in this way have been attempted.
Notably, the Major administration was assessed through the content (and
absences) of the 1995 rural White Paper (Murdoch, 1997a; Lowe, 1996;
Hodge, 1996) and again in the rural White Paper of 2000, which make
little reference to structural issues that are wrought upon the countryside
by globalisation and multinational companies – except to look towards
further reform of EU agricultural policy.

Emphasis instead is placed on an 'inclusive countryside' and measures to
improve the standard of the range of public services in rural areas. In terms
of land-use planning a rather feeble attempt at relaxing controls on diversi-
fication in rural areas is included, an attempt that is unlikely to address

underlying issues of dysfunctional rural 'communities' and more remote rural areas that lack realistic prospects of economic development under such policies (Lowe, 2001; DETR, 2000a). It is the urban White Paper issued shortly before the rural White Paper for England that illustrates some bigger issues – that is to say, an emphasis on brownfield development (DETR, 2000b) that will inevitably play into the hands of restrictive planning regimes and preservationist groups interested in greenfield amenity, who tend to wish to resist any development in rural areas. Rather, citizenship needs networks and the sustenance of elusive 'social capital' that is often cited as the component underpinning rural society and economy.

The advent of an increasingly diverse, plural society brings particular problems for national government as it attempts to square circles and appeal to multiple communities of interest, especially when, as explored above, the individual looks below, beyond and alternatively from the state in order to pursue a range of (self- and community-regarding) interests. The stake, as employed in the Blairite rhetoric, is the interest that the participants hold in the outcome of an enterprise (Ackerman and Alstott, 1999; Macintyre, 1999). It is what the citizen brings to and trades as part of the social contract. One consequence of the postmodernisation, and a post-rural analysis, of the countryside is that benefits and rights feel different and are experienced differentially by different groups. Individuals similarly are differentially empowered with the ability to take up entitlements and the ability to make use of rights. Groups operate on the level of subcultures or thought-worlds that may not be well understood by the state (or the logic of the market) and may not adapt to contractualisation in the way that the state might wish. Rather, individuals seek alternative methods and forge networks that do not respect boundaries in spatial, political or even legal terms.

Reviewing the citizenship projects of recent UK political administrations makes it apparent that politicians have viewed the label as a convenience. Plant (1994) notes that Labour have settled on citizenship as a philosophy and a rhetoric as it provides them with a framework for organising and deciding what is a legitimate claim and what is not. Politicians have also calculated that the label rests lodged in the national, even international, psyche as a positive attribute of status. However, in a multicultural and otherwise diverse society (or perhaps societies/sociations; Urry, 2000), citizenship and identity and calls to participate are deployed almost as fig leafs for political and bureaucratic systems that do not and possibly cannot cope with the full implications of diversity (the interpretation of the Human Rights Act 1998 illuminates this argument well; see Parker, 2001). Such systems similarly cannot evince their ability to deal equitably with interests holding differential power.

Citizenship is, however, a powerful and mutable concept in other senses. Clarke (1996) underlines how citizenship can be dangerous for governments as once the parameters of decision-making and the application of social rights (for example, the right to protest, the right to claim more rights; see

Marshall and Bottomore, 1992: 8) are understood, challenges to the boundaries of legitimacy will be made constantly, and the status quo be subject to a vigorous defence as many citizen rights imply the loss of a feature of another's property.

Rather than drawing societies together, such moves encourage the emergence of multiple sociations based on interest and based on difference and inequality. Smith (1989) claims that this has always been the case: the challenge has been to construct a citizenship that recognises difference and draws people together. Perhaps that notion is also beyond retrieval. This objective has dogged the history of nationhood and of citizenship. It is far from simple for states to reach that degree of inclusivity or flexibility. It is also increasingly apparent that citizens can and will circumvent the state in order to register political views. Hence people are less likely to engage directly with formal national politics, seeking instead to rely on interest groups, conjunctural action, market-based protest and direct action; these examples are illustrated in later chapters.

Citizenship policy is a reflection of power and, more recently, part of nation-states' attempts to regulate and stabilise the social (and cultural) as part of the flows of political and economic change that are created *inter alia* by the effects of globalisation. Conversely, citizenship is also a reflection of agency and the ability and willingness of people to participate in formal and informal elements of governance. Giddens (1995, 1998) claims that governments reflect the values of the society that they represent. Given that society is developing new forms and logics at the scales at which governance is operationalised, the flows and networks between actors can be seen as increasingly important, but Giddens's statement is also to be contested. It is unclear that governments reflect true diversity, rather than predominantly serve powerful and vocal interests. There is increased reflexivity on the part of individuals, but a reluctance still, on the part of national government, to adopt structures that can incorporate social and cultural change in the global age. This reluctance perhaps reflects Gramsci's (1971) comment about state attempts to stem crises in an *ad hoc* fashion rather than seeking to revolutionise practice.

There has been increasing use of discourses that compete with or seek to supplant and subvert national government discourse. Increasingly, supranational forces have frustrated national government projects and policies. Examples include the actions of multinational companies or global action groups such as Greenpeace, or other international movements and political structures such as the European Union or OPEC (see, for example, Beck, 1998; Albrow, 1996; Lash and Urry, 1994). Changes in political awareness from below have also been influencing the ability of government to control political events. Similarly, sub-national governance has served to lead or 'perform' government (Albrow, 1996). A wide range of groups and institutions are appropriating control and initiation of socio-economic change.

In what Beck has termed the internalisation of freedom, a flipside perhaps to the governmentality invoked by Foucault, people are increasingly freed from old social and cultural ties. In the rural context Ray (1999) picks up on this, noting how, in particular, young people brought up in the era of the New Right (so-called 'Thatcher's children'; see Pilcher and Wagg, 1996) choose to exercise their freedom. They tend not to do so in terms of traditionally conceived political and participative citizenship; rather, they form new allegiances based on globalised and localised interests, or based on the realisation of shared, personal or global goals (Clarke, 1996).

Citizenship and glocalisation

The guarantee of our 'citizenship' rights comes through the social contract into which we enter with the state under a system of representative democracy. Yet such a contract is itself subject to continual amendment, especially, one would argue, since power, influence and authority are increasingly 'located' beyond the state. The contract is more political than social as it is predicated less and less on a punctural mandate. Increasingly, the actions and agendas of national-level government have begun to be discussed in reference to supra-national or global governance (Giddens, 1998). Additionally, the nature and generative origins of rights-claims are extensifying, both as a reaction to globalising tendencies and as a response to local, ethnic and historical identities. For example, consider the self-determination pursued by states of the former Yugoslavia, and indeed the claims of the Scots and Welsh, and the English regions, such as the Cornish claims to an independent Kernow, already promoted as 'a region of the European Union' (Cornwall Society, 2000).

In this context it is argued that both the micro/local and the macro/global are impinging on the power and stability of the state. That is not to say that the local is always the micro and that the macro is always the global. One of the fascinating developments of globalisation, more specifically of glocalisation, is the impact that distant local and distant micro can have on other jurisdictions and cultures and therefore on citizenship and claim-rights. Murdoch (1997), in reflecting on the Majorite approach in designing local community involvement in rural affairs, identifies this tendency, linking 'government at a distance' to the breakdown of national-level government and perhaps to the fragmentation of citizenship itself. Rather than leading to global citizenship (Albrow, 1996; Isin and Wood, 1999), this perhaps leads to what can be termed 'glocalised citizenship', whereby citizenship is controlled by a mix of the state, supra-state forces, the local actions of groups and the actions of individuals themselves 'creating their own spaces of control' (Giddens, 1991). The impact and nature of globalisation were mentioned and a discussion of citizenships that are not entirely dependent or formulated at the level of the nation-state was set out in Chapter 2. Here, how such macro-level changes are reflected, or fail to be reflected, by national-level government is discussed.

Counter-urbanisation (as has urbanisation) is playing an important part in reformulating social and cultural dynamics of rural areas (see Boyle and Halfacree, 1998). Pred (1984) importantly makes the point that as well as citizenship being processual, places, both in terms of landscape and in terms of human relations, are always shifting: they are contingent. In his terms they are always 'becoming'. Such a view is uncomfortable for government. Instead, administrations prefer to view the rural and the rural 'community' in terms of geographic and transparent space, in the same way that 'nation' has been understood by the state as the area of land (and sea) under its jurisdiction. Consequently, attempts to construct a unifying discourse to stabilise localities, groups and cultures were needed. For example, any such unifying discourse was, of course, found to be wanting in India, Africa and other outposts of the British Imperial past and thus a new unifying concept of 'Empire' was required in order to help draw people and place together (see Mol and Law, 1994; Murdoch, 1997a).

An outcome of 'the reflexive project of the self' as posited by Giddens (1991) and of the new sociations identified by Hetherington (1996) is the reorganisation of the social into single-interest, often time-specific, 'communities' or affinity groups. Such communities of interest are not necessarily fixed or homogeneously located in space – although they may make claims over space for their own purposes (e.g. the Ramblers' Association) – as opposed to traditional communities defined by location and boundary (Crouch and Matless, 1996). They are an important aspect that has tended to be overlooked in traditional studies of rural politics. In looking to protect 'community' there has been an unwillingness, especially on the part of policy-makers and politicians, to engage with the forces and processes of the global and the postmodern. It is increasingly clear that a particular formula of social glue that used to hold groups together in rural areas is being replaced with new dimensions and dynamics of self-interest, for example the amenity value and experience/consumption of (British) landscape demanded by (glocal) citizens.

There are disruptive, often complex, cross-cutting interests of different people who may be located physically in the countryside, but to whom traditional conceptualisations of what it means to be 'rural' hold little relevance and to whom the imagined rural is a far cry from the reality of their surroundings (see Savage et al., 1992; Cloke and Little, 1997). It is clear that citizenship is about a contract that has multiple attendancy. A citizenship, in terms of state construction, for the post-rural cannot be forged if the rural is still conceptualised as it has been in the past, where homogeneity, stability and stewardship by the few remain dominant. Conceptualisations of citizenship by government retain a structured narrowness; however, citizenship is being undertaken proactively already – citizens perform the state, and if anything, certain citizen groups representing certain interests are becoming more active. There is a challenge for governments that wish to serve all interests, though. Those administrations that wish to mobilise citizenship as core parts of their political projects should make an honest and determined effort to enable and foster true, perhaps uncomfortable, active citizenships across claims based on class, interest, issue and from other figurative groups.

How, then, might citizenship be reconceptualised for a globalised, post-modern era? How should government seek to envision citizenship and rural regulation in a way that is conducive both to existing rural people and to those wishing to use, or in some other way be involved with, rural areas? Possible ways include iterative participation in conservation, concern over sustainability, food quality issues or amenity use as well as direct involvement with traditional industries or users of services.

Conclusion: alternative agendas and citizenship

Citizenship, as noted, is continually being contested, challenged and amended from within localities by individuals and groups; by interests without fixed or place-specific claims; from the distant local, the global; and through the politically inspired restructuring of national government.

When a reworked citizenship theory is applied to interests attempting to remodel citizenship and the social contract vis-à-vis land and rural space, some interesting insights become apparent. Recent developments illustrate how the rural has been portrayed by different interests as being 'in crisis' (Countryside Alliance, 1999; Barnett and Scruton, 1999). Once such a stance or problematisation has been adopted, solutions are necessitated. Identifying the same problem, however, does not imply that the same priorities or the same solutions will be ascribed to the problem. It is also the case that other, overlapping problems may be identified: different frames of reference, different epistemologies and different definitions of rural may impact on the already recognised problem and thus prompt further, different solutions – for example, in terms of the scale and depth of required change.

Global-level shifts and developments that have impacted on different geographical scales and in social terms have reworked relationships between different people and different, often distant, places. The issue of the conversion of rights to and from the public domain and across different scales of governance and decision-making are aspects of a struggle for power. The rights debate can be couched in terms of freedom and equality and the role of the state, the market and the citizen. Citizen self-help, the penetration of the market discourse into the countryside, and of government and interest-group actions to preserve or claim those interests as rights in the countryside are burgeoning, in order both to argue for change and to resist change.

British politicians have adopted a conservative view in rural affairs in the post-war period. Unfortunately, such conservatism is likely to be continued under the Blair administration, even though the prime minister has been at pains to claim that he is committed to fighting the 'forces of conservatism'

(Blair, 1999). For both the main political parties the rural constituencies provide the key to political power. For Labour this is a new experience, and one consequence of their new-found hold on the shires is a reluctance to upset them. Over the past twenty years successive governments have defined citizenship narrowly and sought to maintain a range of status quos in the countryside. Even though there has been change, very little has been radical – rather, change has been aimed at the amelioration of underlying issues (see Winter, 1996). While New Labour have calculated that issues such as countryside access and hunting can be tackled and the party can still maintain its popular support, they are very wary of more radical measures involving, for example, land reform, planning, and tackling the European Union and big business on food and other production and service issues. Unfortunately, such a situation leads to 'more of the same' for the countryside, a policy that is increasingly untenable for small farmers, traditional communities and those without cars or ample incomes. Thus the policies being pursued and issues left largely unchecked make for an inconsistency that negates many of the attempts to redress rural ills in documents such as the 2000 rural White Paper.

Instead, government has attempted to shore up extant lifestyle or thought-world conceptions of dominant groups. These attempts have led to increasing conflict in the countryside (and elsewhere) and an increase in demands from marginalised, sectional or urban-based interests that their claim-rights be met or maintained by the state, even when those patently cannot be ensured at the national level of governance or inherently conflict with existing governmental arrangements.

There has been little willingness yet to engineer fuller political citizenship where people volunteer not only their time and energy for 'good causes' but are enabled and encouraged to practise, and are honoured for doing so, the day-to-day exercise of political consciousness. Crises presented or constructed as being from within or outside (often threats upon the state itself) have required the state to treat on different terms with factions of society; some become less important while others gain in political value. Citizenship as propounded by government has been partial, and government has, by and large, expected citizens to be part of a central, national 'project' or state-defined social contract in which the role of the citizen is largely one of recipient and passive provider of legitimation for politicians and policy-makers. However, Turner also recognises that citizenship incorporation can imply a very different type of citizenship:

> where citizenship develops from below (as a consequence of class struggle) then we have an active and radical form, but where citizenship is imposed from above as a 'ruling-class strategy' of incorporation, then we have a passive type of citizenship. In addition, where the public sphere is not regarded as an appropriate moral arena (for example, where the family is seen to be the 'natural' space for the moral development

of the citizen), then politics becomes privatised, reinforcing the passive nature of citizenship.

(Turner, 1994: 159)

A clear example in this respect has been the preoccupation of the Institute for Citizenship Studies – formerly the Institute for Citizenship – with national education for normative citizenship (see Institute for Citizenship, 1992; Institute for Citizenship Studies, 1999). In the course of the book, citizenship in its normative sense is deliberately attenuated, especially when assessing the ways in which people engage with or participate in society. This includes the rise of DIY culture (see McKay, 1998; Halfacree, 1999) and the appropriation of the market as political space (see Shotter, 1993; Urry, 1995; Parker, 1999c; Klein, 2000, as discussed in later chapters).

4 On being modern

Consolidating citizenship in the countryside

Introduction: ordering the countryside, ordering citizens?

Like citizenship, dominant conceptions of land use, ownership and management are currently under criticism, and historical examination can be an
important contextual method in reassessing present structures. Further than
this, the land and space form, and are formulated by, citizenship constructions. If, as suggested earlier, citizenship is a reflection of power, then land
and its use can be viewed similarly. This chapter provides a particular and
partial social history of rural England with a focus on the way in which the
countryside, and in particular land, has been administered. The chapter uses
aspects of rural history to combine a review of the development of citizenship with a discussion of the development of land rights structures in the
UK, also noting how current debates over land use are being informed by
the historical. The primary justification for this part of the work is that the
historical formulation of rights distributions in the countryside is placed
centrally, charting the historical and territorial legacy that this now provides.
It is also argued that history is being constantly and partially deployed by
various interests as a mobilising or legitimising discourse (see Billinge, 1993;
McKian, 1995; Parker and Wragg, 1999). In this way an historical narrative within which to assess citizenship in the countryside is provided.
Mobilisations of history and heritage in the countryside are further discussed
in Chapters 5 and 7.

The methods and processes that shape and influence the nature of rights
are varied. Often it is through public pressure, perhaps protest, that socio-
political change is asserted. A recent example of state intervention in shaping
the way in which such protest can take place is the introduction of the
Criminal Justice and Public Order Act 1994, which, among other contro-
versial effects on rights, criminalises trespass. The historical development and
curtailment of rights in the English countryside forms an important connection between land and rural space and citizenship and is detailed in order
to outline how land (and place/space) was organised and structured. In
similar fashion, modern citizenship required both uniformity of rights and

duties and the conformity of people in order for it to fulfil its part of the modern social contract. This, as Colin Ward states, represented the 'officialising' of the countryside (1999: 190), whereby unofficial uses and users were guided towards state-defined behaviours or alternative physical locations – that is, the towns. In another sense, then, reviewing moments and processes of controlling space through time is important in connecting citizenship rights and land rights.

Contemporary protest over land and protest/participation in the political process of gaining access to land for recreation is another of the threads of the book. Historically, the issue of public access to land has provided examples of public protest and attempted 'participation' in predefined political processes. In order to make the task of detailing such historical change more manageable, the prism of land rights and associated rights (for example, use rights and political rights) is used to highlight the central thesis of contingent citizenship and the conflicts associated with social and cultural aspects of 'rights' trades and transfers. This approach inevitably means that some aspects of the historical construction of (and resistance to) citizenship and rights in the countryside are somewhat neglected. It is a complex, interconnected and diverse mix of concerns. The primary focus is on the dual development and apportionment of rights and duties over land and over people rather than the specific inadequacies of that situation or on other aspects of citizenship envelopes, such as health, education, and minority rights.

Many of the 'rights' discussed in relation to land were historically customary (or *de facto*). Thompson (1993) observes that customary rights were rarely formalised, while Hill (1996) notes how customary 'liberties' were observed and developed unevenly and irregularly. The informality of such a situation may be difficult to appreciate fully. Thompson writes that the life of the country-dweller was full of an 'ambience' – a lifeworld based on custom and reciprocity. In order to explain this, Thompson invokes Bourdieu's notion of habitus as 'a lived environment comprised of practices, inherited expectations, rules which both determined limits to usages and disclosed possibilities, norms and sanctions both of law and neighbourhood pressures' (Thompson, 1993: 102). Aspects of Bourdieu's cultural theory are also threaded through subsequent chapters, helping to explain the economic, political and social shifts that citizenship has undergone and how new forms of citizenship are developing. It is argued that this has significant bearing on the development of rights and citizenship in relation to land use and, more widely, to the attachment to an imagined rural with imagined and enforceable rights. In one sense, rights can be seen as an expression of the habitus and the field of power in which individuals exist (see Bourdieu, 1977, 1990; Jenkins, 1992; Isin and Wood, 1999), as is resistance to extant rights *qua* definitions of appropriate practice.

Habitus and social field are important conceptual markers in this analysis. The habitus is not only an historical notion; habitus is contemporaneous, altering with political, economic and cultural change (as well as being susceptible to influence by others, perhaps distant others' changing under-standings or reflexivities). As such, the habitus defines part of the social and economic position of the citizen (Warde, 1994) and implies a degree of durability in citizens' understanding of their 'proper' relationship to different spaces and places. It could be said to represent the lifeworld of the indi-vidual, acting as structuring agent for the citizenship 'envelope' (in the widest sense) present in any particular locality. The lifeworld relates in part to the notion of 'practical consciousness', which Giddens (1984) explains as the tacit knowledge or 'cultural competence' (Cloke and Little, 1997; Eder, 1993) that may structure individual identity and lifestyle. Such consciousness allows the individual to undertake routines, activities and interactions without prior reflection or preparation. His or her actions are, in these situations, dictated by the habitus. Rights and responsibilities are understood as being reflections of structures that shape the habitus and as being justified under the conditions of the field of power.

A brief overview of how various rights and cultural expressions in rural areas were altered from the English Civil War onwards is set out. The chapter incorporates important events and processes such as the enclosures and the claims of the Diggers in terms of rights in order to illuminate the wider thesis. The date from which such a review starts is inevitably rather arbi-trary. However, the justification for limiting the review in this way is that the formulation of an explicit social contract was coming into being from that period (Lessnoff, 1986; Hill, 1996). This involved land and artefacts, practices on land or features in the landscape being appropriated, moved and rationalised, with traditional social and cultural arrangements being sacrificed or traded for other benefits, or aspects of the contract.

The 'dustheap' as reservoir

Hill lucidly argues that the landless, the 'outlaws, beggars, the poor, vagabonds . . . godly nonconformists' (1996: 325) were the objectors to the law in the seventeenth and eighteenth centuries. It was they who resisted the modernising hand of the powerful minority, especially as it applied to the control of space, and specifically land and its use. This aspect of rural social history is important for a number of reasons, and a reading of the historical acts as link between the rationalisation of land in the past and the more contemporary claims over land and to citizenship in the countryside.

Representations of past events and justifications for past actions are power-ful; texts that recount the past represent congealed power (see Murdoch, 1997a; Latour, 1994). Increasingly, interest in the subject of 'lost' and ancient rights over land has led to recent protests and claims over such rights to be restituted, prompting calls for a re-examination of present rights distributions and contemporary land management practices (see, for example, Shoard, 1987, 1999). There has been a shift from the modern to the postmodern era to produce revised and alternative versions of history. The challenging and appropriating of historical representation can be a very useful political tool, especially for disputing extant structures and norms in society or indeed in drawing on history and 'heritage' to defend status quo positions. Past policy and the historical legacy still help to shape the ways in which future countryside policy may develop. The conditions of possibility may change but the historical aspect, the prevailing history, sets the country-side on a particular trajectory. Bender (1993: 275) notes that this importantly involves 'mobilising different histories' and that such histories are 'differen-tially empowered through time'. Examples are used in Chapter 5 to link this historical element to the contemporary, where it is shown that old 'reservoirs' of power were drawn upon in an attempt to destabilise dominant conceptions of appropriate land rights, planning priorities and natural resource uses.

This chapter makes an important link between the development of national citizenship and the rationalisation of land and its use under a capi-talist system. The development of the state in Britain (though admittedly the work has an anglocentric bias) provided the institutional shell within which rights and duties were developed; indeed, formal citizenship is not possible without a form of state structure (see Bromley, 1991; Jessop, 1990; Poulantzas, 1968). Hence the organisation of rights is reflected in the way that land is regarded legally and culturally, in line with Lefebvre's (1991) position that producing space is homologous to the production of citizens. Giddens (1985: 203) observes that

> only since the eighteenth century have the three traditional strands of citizenship rights (civil, political, social) become distinct from one another. This is partly because each has a different organisational focus or, at least, the first two [civil, political] do. The main institutional focus of the administration of civil rights is the legal system. Political citizen-ship rights have as their focal points the institutions of parliament and local government.

The agricultural 'revolution' and the redistribution of rights

Common rights and the trades in rights that took place in the period of the 'long revolution' in the countryside are a good example to help provide an introduction into the theory of rights over land. More generally this example illustrates how citizenship development and rights (re)distributions were important elements of socio-economic change in rural England during the period of enclosure, primarily during the seventeenth, eighteenth and first half of the nineteenth centuries (see, for example, Thirsk, 1967; Trevelyan, 1967; Beckett, 1990; Thompson, 1993; Hill, 1996). The process of improving and ordering the land and the landscape was also, concurrently, taking place over already 'privately' owned land, and similarly for universal norms of acceptable use and behaviour with regard to land-developing.

In the eighteenth century, for the first time attempts were made to apply law evenly and abstractly by 'the state' – a shift from the 'gamekeeper' to the 'gardening' state (Bauman, 1987). The law was applied to all in a territorially delimited area and became increasingly more universalistic, 'helping to de-personalise state authority from the ruler, thus providing a basis for abstract rights and duties to be conceived separately from the cumulated prerogatives and rights of the historically distinct status groups of feudal society' (Clegg, 1989: 251). In terms of legal regulation the law was being created and used in order to rule, rather than as a mutually agreed code of conduct. The enclosures were a key example in this respect and are duly discussed below. When this use of the law in order to rule is viewed alongside repeated exhortation on the part of successive governments (as seen in Chapter 5) towards citizenship and rights/responsibility discourse, there is an interesting connection to be made between this historical development and contemporary strategies towards 'managing rights'.

The impacts of change throughout history have shaped not only the physical and legal manifestations of rural areas that we see today, but also the cultural mix. All three are linked and provide elements of the political frame within which the study of the countryside is approached here. Some significant events have had radical impacts on the countryside and on the citizenry, consequently altering rights distributions and identities. The main concern here is to analyse the changes in the distribution of rights in the countryside, especially during the period of the Agricultural Revolution, where symptomatic change such as enclosure and the redistribution of land ownership had been occurring gradually and irregularly since the Middle Ages (Douglas, 1976; Hill, 1996; Thirsk, 1999). The process of land rationalisation that took place during the 'enclosures' is discussed as one such key phase. It is noted how altering land rights and therefore 'citizen' rights was closely associated with political and cultural change being promoted by particular classes. Indeed, the governance of land (and space) can be viewed as an important part of the project of modernity (Lefebvre, 1991; Rabinow, 1991). The reformulation of land rights and relationships to the land was part of a series of cultural, legal and physical alterations in train in England particularly, but also in the UK more widely, for at least the past four hundred years. Those changes enabled the state to rationalise the land and population through various mechanisms of governance.

The charting of the development of citizenship and land rights structures (both legal and cultural) cannot proceed justifiably without investigating the linkages between changes in rights, the rise of capitalist structures and the main economic and political events of those periods. A brief and partial history such as this requires that much of the complexity be downplayed in order to allow the main themes of citizen formulation and land rights to remain discernible. The commentary focuses therefore on the consolidation of land and rights through the period of enclosure and the Agricultural Revolution, and then looks at how modern society has tailored rights and responsibilities under capitalism and systems of representative democracy. A section on contemporary challenges to the present situation provides a bridge to examples outlined in subsequent chapters.

It is normally argued that the right to exclusivity of use is cardinal in the bundle of rights that constitute the Lockean–liberal conception of private property rights (Becker, 1977; Denman, 1978; Munton, 1994). The construction of English land law reserves the right of landowners to exclusive use of their land. This conception holds the doctrine of possessive individualism as central to a property-owning society (Honore, 1961). To relinquish this facet of ownership would be a fundamental change of character in the putative private property rights structure. The cultural expression of the dominant and reductionist attitude towards property and land is exemplified in phrases such as 'an Englishman's home is his castle', an attitude that has been systematically reinforced over the past four centuries and is reflected in current policies, practices or legislation such as Neighbourhood Watch and Farm Watch (Yarwood and Gardner, 2000). In recent years there has been very little public debate over land reform issues in England. The land reform movement that grew up towards the end of the nineteenth century under the influence of people such as Henry George largely died out by the end of the First World War even though exhortations to fight for the country were powerful calls to arms (Douglas, 1976). (To be fair, to this day the Henry George Foundation is active in lobbying on land reform issues worldwide.) Gradually, demographic change, the 'need' for productive land and an increasingly urban and industrial population looked towards the rational, technical solution of land use planning presented as the solution for the social reformers, particularly as the rural aesthetic appeared under threat from unchecked urban sprawl and unfashionably organic 'plotland' landscapes (Hall and Ward, 1998). Ebenezer Howard, among others, shifted the emphasis away from a restructured rural demanded by Victorian notables such as Joseph Chamberlain with his smallholding vision of 'three acres and a cow' for all who should want it, instead pointing the way to the good life with his garden cities prospectus published in 1898 and a vision of a reclaimed, social urban (Hall and Ward, 1998; Cherry and Rogers, 1996).

The circumstances under which dominant interpretations of rights-claims are challenged are intensely political. Where and how certain rights are rolled back and new ones introduced as social, historical, economic or political circumstances alter, is problematic. The emphasis that the judiciary give, in terms of citizens' rights, to the protection of private property rights (over social/cultural rights) is part of the historical development of the legal framework. This means that individual claims are often protected or favoured over or against general or group claims. The interpretation and support of the judiciary has been crucial in developing and then maintaining a certain conception (and therefore distribution) of citizenship rights. When this conception becomes outmoded, then the development of a distribution of rights that reflect the most equitable outcome for the whole of society becomes a responsibility of government (Becker, 1977; Bonyhady, 1987). It is a fairly well-rehearsed argument that historically this democratic, social obligation has been obscured by the politics of self-interest and possession, and less well recognised that change is blocked, subject to compromise or effected tardily.

Enclosing land, enclosing citizenship

During the period from Elizabeth I to Queen Victoria and beyond, the combination of urbanisation, commercialisation, the imposition of quasi-religious social values and the rationalisation of land use led to a significant shift in customary *de facto* and legal *de jure* rights (Malcolmson, 1973; Beckett, 1990; Tawney, 1926). Such rights in retrospect can be viewed simultaneously as both land rights and citizenship rights. These were proscribed legislatively, bought out, or created as part of a process of social conditioning or modern governmentality (Poulantzas, 1968; Jessop, 1990). Land use was controlled in order to bring about suitable conditions for capital accumulation and the development of agriculture into a profitable enterprise, and so that ownership could be agglomerated with as little hindrance as possible. The consequent gathering of rights that accrued to landowners is crucial to present-day conceptions of 'landownership'. The power that landowners then clinched, using the land, is important in analysing present policy constraints.

The effects of the Agricultural Revolution on rural society were profound and manifold, and the repercussions for rural society in contemporary times are still quite fundamental. Newby (1987: 8) emphasises this point when introducing the Agricultural Revolution as a 'transition between two completely different types of society . . . it is a question of the whole social, economic and cultural basis upon which society was constituted'. The enclosures in particular embodied the ongoing process of change during the period of the Agricultural Revolution. In terms of the rationalisation of rights and the formalisation of such rights over land, the period was radical (Trevelyan, 1967). The changes outlined below were brought about not by the expansion of the old economic system, but by the creation of a new one. The implications of the move away from a feudal system to a market-based system were far-reaching. The way in which the land was used changed, and as a direct consequence the way of life of many rural people changed. The impacts and outcomes were economic, cultural and political. Therefore the Agricultural Revolution importantly shifted the trajectories of rights and identity.

The features of the transition from a feudal countryside to a capitalist one were varied, the enclosure movement having significant roles, both pro-active and reactive, in facilitating such dramatic change in rural land use. The economy and the population were both in a relative state of flux. There was change in methods of production and technological advancement; there were large increases in population, with the process of urbanisation being a feature of their distribution (see Beckett, 1990; Hilton, 1976). Such periodic rises in population did also lend weight to attempts to further rationalise land use to make land more productive (see Thirsk, 1999).

Between 1650 and 1850 there was an increase of approximately 300 per cent in the population of England and Wales. By 1801 the population stood at around nine million and was rising, with the birth rate being far higher than the death rate (Trevelyan, 1967; Turner, 1980). These increases in population, which continued throughout the Victorian era, served as part of the impetus for increasing the productivity of the land. Couched in simple terms, the more people, the more that pressure would grow on the land for a variety of land-use activities, and the more the land would be valued in both social and economic terms. Therefore the power and prestige that had already been a feature of landownership became even more important as a means of capital accumulation and as a source of power (Thompson, 1973, 1975).

The transference of rights over land (as citizenship rights) to rights in land (as private property rights) over time has importance for identifying policy areas that reinforce or depart from the particular course of development which citizenship rights and private property rights have taken since the Agricultural Revolution. The historical legacy of the Agricultural Revolution and the features of the process of change in rural land use are ones that saw small numbers of people (rights-holders essentially) having 'traded off' potentially flexible (and communal) land use. There are numerous tenets of private property ownership that are commonly identified as constituting the 'bundle of rights' often referred to in texts concerning property (see, for example, Bromley and Hodge, 1990; Honore, 1961: 112 on liberal ownership criteria). Bromley (1991) argues that these over time have become firmly embedded and legitimised. It is argued that the Agricultural Revolution and the enclosures in particular signify crucial elements of the ordering of space to enable more suitable conditions for social control. This allowed a governmentality and a state-making to be nurtured that prioritised individual, exclusive entitlements (to property) and attendant obligations to observe them rather than coexistent cultural reciprocities (see Mann, 1987; Simons, 1995; Jones, 1999).

Enclosures

The enclosures were part of a wider socio-economic shift towards the rationalisation and economic utilisation of the land. This series of significant changes included increased capitalisation of the land, improved farming techniques and the development of important agricultural innovations (Newby, 1987). Before the advent of widespread enclosure, market-oriented land use and the 'drift from the land', country life had been dominated by the ancient feudal system whereby rights of ownership were less closely defined or enforced than at present. 'Ownership' was less concentrated in private hands, since rights to and interests in the same land parcels accrued to many people. The shift into modern unified rights of ownership was a gradual and uneven process spatially and temporally, bringing with it uneven changes in the observance of universal rights and obligations (Hilton, 1976; Hill, 1996).

Approximately one-fifth of the land area of England was enclosed between 1700 and 1850 (Turner, 1980; Thirsk, 1967; Cherry and Rogers, 1996). This amounted to two and a half million acres of common and waste and four and a half million acres of open field, with many other parcels being enclosed before and after this period (Williams, 1973; Hilton, 1976; Turner, 1980). The enclosures involved the demarcation of land from common open land, either uncultivated or land farmed on the open strip system, to clearly defined, and often larger, parcels of privately owned land. This involved the allocation of land to individual ownership or occupancy. Much arable land being farmed using the traditional strip system was enclosed. The land that was enclosed was usually fenced, or bounded by ditch or hedge, to signify its enclosure. Accordingly, many of the features in the current rural land-scape represent artefacts of private power and private mastery of rural space (and place). Thirsk (1967: 125) states that:

> To enclose land was to extinguish common rights over it. . . . To make it economically worthwhile, enclosure was often preceded by the amal-gamation of several strips by exchange or purchase. If the enclosed land lay in the common arable fields or in the meadows, the encloser now had complete freedom to do what [s]he pleased with [her]his land throughout the year.

Although the enclosures were only one feature of the Agricultural Revolution, they were 'possibly the most important, of the many changes that combined to reduce the numbers of the independent peasantry, while increasing the aggregate wealth of the countryside' (Trevelyan, 1967: 390). It has been suggested that one of the prompts for enclosure lay in 'an altru-istic desire to feed a growing population' (Norton-Taylor, 1982: 17; see also Thompson, 1993), a claim that might easily be made today for a very different regime (see, for example, Fairlie, 1996; Bromley and Hodge, 1990). Indeed, on that line it was said to be the small farmers, commoners and squatters who lost most from the original process and who have continued to suffer under a 'modernist' approach (Chambers and Mingay, 1966). Some enclosures were carried out with the active collaboration of the peasants themselves. Others, especially the enclosure of commons, 'were deeply resented, and provoked riot and rebellion' (*ibid.*: 131).

The later phase of enclosure was truly state sponsored. Around four thou-sand acts of Parliament were passed enabling enclosure (Cherry and Rogers, 1996). In order to bring a bill of enclosure to Parliament a majority of the landowning interests of the land in question would need to agree to its enclosure. There were many enabling acts of Parliament that altered the eligibility of applications, notably the 1801 General Act, where the consent of two-thirds of the landowners was needed to allow an Enclosure Bill to proceed. It was normal practice for larger landowners to buy out smaller landowners specifically to get the necessary legal consent to enclose (Turner, 1980). Between 1720 and 1850 there was a relative spate of enclosure, mainly enabled via parliamentary bill (see Chambers and Mingay, 1966). The process of enclosure had by this time become national policy. Trevelyan (1967: 391) notes that

> after the third decade of the eighteenth century the work began to be carried on by a new and wholesale procedure: private acts of Parliament were passed which over-rode the resistance of individual proprietors to enclosure; each had to be content with the land or the money compensation awarded to him by parliamentary commissioners whose decisions had the force of law.

The parliamentary enclosures amounted to large-scale compulsory purchase of a wholly unsophisticated kind. Methods of valuation and compensation were somewhat crude and the appraisal of such esoteric things as common rights or customs were inevitably considered vexatious by the valuers. In that sense, local culture was something to be eradicated rather than valued. As might be expected, arguments regarding the intrinsic value and potential value of such customary rights were not heeded at the time (Hill, 1996). The loss to future generations was not a consideration that could sufficiently oppose the economic arguments for untrammelled private property rights (Thirsk, 1967). The political motivations of the enclosers during this phase of enclosure are not difficult to detect (*ibid.*: 391): 'Batches of these revolutionary Acts were hurried through every Parliament of George III (1760–1820), assemblies not otherwise famous for radical legislation. But this was the radicalism of the rich, often at the expense of the poor.'

We can see from the historical development of common rights that the theory of law, developed to insist upon individual ownership, was instrumental in enabling land rationalisation and therefore of rights to proceed 'legitimately'. The theory of law had considerable influence over the way that rights were conceptualised; thus the state indirectly regulated those rights which had originally evolved and been passed down informally since before the development of the modern state. Thompson reiterates the point that enclosures represented an opportunity for a few fortunate ones to 'cash in' and monetise rights: 'it signal[led] a wholesale transformation of agrarian practices, in which ancient feudal title [was] richly compensated in its translation into capitalist property right' (1993: 137). Thus use and users were gradually separated on the grounds of improvement and so that property rights resided with the individual (see Bromley and Hodge, 1990; Bromley, 1998).

Rights and enclosure

The theories of John Locke and, later, Adam Smith gained increasing acceptance in the seventeenth and eighteenth centuries (Hodge, 1991; Bromley, 1998) and were used by property-owners to justify the exclusion of others via the laws of property as laid down by a Parliament that was itself dominated by the landowning interest (Norton-Taylor, 1982; Hill, 1996). These political and economic theories emphasised and gave priority to the atomistic and the contractual – the doctrine of possessive individualism (Honore, 1961; MacPherson, 1962). It is true that the feudal economy was, in normative terms, inefficient and lacked order. The new order was 'conforming with an age of agricultural "improvement" and was finding claims to coincident use rights to be untidy. So also did the modernising administrative mind' (Thompson, 1993: 106). The philosophical reinforcement of the 'modernisers' and conceptual/ethical justifications for change were developed such that 'Britain's landowners acquired an ideological framework of their own to match that of their critics . . . they looked to philosophers who presented the ownership of property as intrinsically good' (Shoard, 1987: 60).

The later period of enclosure took place within the wider backcloth of the Agricultural Revolution and the beginnings of the Industrial Revolution (Trevelyan, 1967). An increase in the supply and productivity of agricultural land was one of the prerequisites of the former and an increasingly urban-based workforce a result of the latter. Enclosure at this time was therefore an important push factor in the nascent modernising industrial age. A detailed consideration of all the factors driving change in the eighteenth and early nineteenth centuries is not possible here (for this, see Thirsk, 1999; Winter, 1996). There were, however, several main conditions which precipitated further enclosure of land, notably the sharp increase in the prices of grain crops, prompted in turn by the Napoleonic Wars and the Corn Laws (Trevelyan, 1967; Turner, 1980; Chambers and Mingay, 1966). Indeed, such projects were reinforcing the project of state-building and the 'normalisation' of relations based on the new legal order. These are also clear examples where events deriving from circumstances beyond rural England were already driving significant changes in land and citizenship rights (not least the Irish land question, which gained in political significance after the famine of the 1840s; see Douglas, 1976).

Alterations in the structure and distribution of rights throughout the periods in question gave rise to ill feeling among the local populations who were to lose 'rights' under the rationalised landscape of private property. The expedited enclosures of the period 1760–1820 made the changes more readily apparent to the population at large. Thomas Paine viewed the attachment of rights to place rather than to person as being rather absurd. Now, however, we seek to attach 'rights' to all manner of non-human and abstract entities. Rights of access to the countryside for recreational purposes are a classic example of this (see Shoard, 1987; Parker, 1997). The conflict and overcrowding of commons by users inevitably led to the disaggregation of use rights from the user (Thompson, 1993: 137). The law in the enclosure period was used as a tool of class expropriation. Rights were quite determinedly reorganised to favour production and larger owners. Those who stood to benefit most from land consolidation constructed the justifications. Rights were increasingly viewed as being attached to ownership of property/place or as associated with standing within the community, essentially being conditional on a number of legal, political and cultural factors.

The cultural role of land and its ownership changed. A discourse of stewardship was in its infancy but was being mobilised to support the consolidation of new entitlements (Woods, 1997; Parker and Ravenscroft, 1999). The intensive use of land as a wealth creator became of paramount importance. The courts applied the laws relating to private ownership rights, such as exclusivity of use, vigorously. The outcome, which characterises all the enclosures, regardless of their original motivation or intent, is the homogenisation of land into relatively marketable and ultimately profitable parcels, which could be used variously and without interference from minor interests in the land. The land had become a commodity to use and to trade, and customary practices and common rights at that time were not in the interests of those holding power. The modernising (but hardly invisible) 'hand' of Parliament and larger landowners (not necessarily mutually exclusive groups) saw such 'rights' as hindrances to the rationalisation of land and its use (Thompson, 1993; Hill, 1996).

The extensive nature of common rights and widespread ownership of small packages of land ameliorated the meagre lot of the peasantry. Rural land was either waste, common or under cultivation using the open field system, whereby different strips were owned by the individual on a subsistence basis (Trevelyan, 1967). The later capitalisation of the land and improvements in farming techniques were to make agriculture an activity that could create surplus produce and therefore be profitable. The notional economy of coincidental use rights was coming under greater strain, as Thompson (1993: 106) notes:

> Demographic pressure, together with the growth of by-employments, had made the marginal benefits of Turbary, Estover etc. of more significance in the package that made up a subsistence economy for 'the poor'; while at the same time the growth of towns and, with this, the growing demand for fuel and building materials enhanced the marketable value of such assets as quarries, gravel and sand pits, peat-bogs, for the larger landlords and lords of the manor.

Gradually the relationship between the landed and the landless changed from cohabitation and the acceptance of traditional and customary relations, to one that relied on law to enforce a particular rights structure. The interest of the common good gave way to private interest (Shoard, 1987). The emphasis on land as a commodity or as strictly private is one which has developed over time: 'private property in land, is itself a concept which has had an historical evolution. The central concept of feudal custom was not that of property but of reciprocal obligations' (Thompson, 1993: 127).

The physical act of enclosure was a manifestation of the gradual change in the distribution of power throughout society and the nature of the exercise of that power. Williams (1973: 107) underlines the importance of the enclosures as a social, economic, cultural, legal and political statement:

> What happened was not so much 'enclosure' – the method – but the more visible establishment of a long developing system, which had taken, and was to take several other forms. The many miles of new fences and walls, the new paper rights, were the formal declaration of where the power now lay. The economic system of landlord, tenant and labourer, which had been extending its hold since the sixteenth century, was now in explicit and assertive control. Community, to survive, had then to change its terms.

The impact of the enclosures should not be underestimated. The effect they had, on a cumulative scale, on many spheres of life was dramatic. The modern economic system brought about a redistribution of rights over land; the loss of customary rights was one of the most important consequential changes and one that has provided the legacy, which directly affects public use of land (for recreational purposes) now.

Traditional 'rights'

The enclosures began the process of codifying rights and consequently altering rural culture (Hoskins, 1963). Shoard (1987: 66) emphasises the effect this had on the poor and those without their own land:

> After enclosure all common rights disappeared [on the enclosed land] except, in some cases, the right to glean fields after harvest. The poor lost their right to graze animals, cut turf, gather wood, collect berries and so on. In countless villages in England and Wales, the effect of the changes was to destroy the subsistence economy that supported the poor.

'Common' rights to fish, take wood and cut peat or to graze animals were important, as were other customary rights – for example, recreation (see Clayden, 1992). Before widespread enclosure and the linked agricultural changes, communities had substantial access to land (Bonyhady, 1987; Malcolmson, 1973). These rights were still rooted in exception and benevolence (see Thompson, 1993; Parker and Ravenscroft, 1999). It also became a bone of contention that many such rights were guarded exclusively to the detriment of the landless (Hill, 1996; Boulton, 1999).

The various common rights to use or take from the land were lost to the peasantry and, with them, extant and future recreational 'rights'. Use rights implied the ancillary of 'access'. It should be said that it is the purpose of the access to land that makes the difference at this time. Later the purpose became less important, as the exclusivity and power of exclusion became an important symbolic right of ownership. Malcolmson (1973: 108) recognises, however, that many enclosures did adversely affect the exercise of popular pastimes:

> Enclosure militated against popular recreation since it involved the imposition of absolute rights of private property on land which had previously been accessible to the people at large, at least during certain seasons of the year, for the exercise of sport and pastimes.

Rural culture and traditional recreations were consequently suppressed, and effectively curtailed by the enforcement of private property rights over land that had previously been used for recreation. This trajectory of rights development carried with it the associated effect of precluding future demands on that space for leisure. Jones (1989: 116) makes this point:

> The economic or rational approach to relationships in the countryside was accompanied by an assault on traditional village culture. Once again the targets were the 'idle', 'dissolute' and 'desperate', and the objectives were control, respectability, and productivity. The attack, which came from both outside and inside the village, was conducted through the Church, the school and the law.

The change to an industrial urban economy and a commercialised rural economy left the role of recreation in a much-altered situation. Access to land was curtailed for many activities previously enjoyed by the landless. Part of this access to land was for recreation, and the opportunity for such recreations was, therefore, restricted, along with the curtailment of hunting and other rights in common. The distinctions between access to land for work, leisure or subsistence were blurred, and the use of land for the purposes of leisure activity was almost certainly not the most important facet of the 'lost rights' to land for common people at that time. The opportunities for recreational use were, however, present and many cultural, recreational events were based on the land, themselves being nascent and customary rights.

Here access to land has double resonance both in terms of the historical and multiple connotation, and also for the contemporary struggle for amenity access to land (DETR, 1999) and a resurgence of calls for access to land for other purposes to be reconsidered (Fairlie, 1996; TLIO, 1998). It is also relevant given the practices, such as foxhunting, that were exercised and growing in popularity among the landed and wealthy, and eased by rationalisation of land use. Hill (1996) notes that the enclosures also had the effect of regulating time such that the peasantry were forced into wage labour and time-constrained labour relations, which consequently restricted their freedom of choice and established the distinction between 'free' time and working time. Such developments had the twofold effect of distancing people from the land and creating a desire to return to it in some way.

It is difficult to distinguish between amenity access rights and rights of access to land for purposes other than recreation, which may have existed before and during the main periods of enclosure. In such altered economic and social conditions the categorisation of use rights that existed historically loses some degree of meaning when viewed in the light of present legal, political and societal contexts. The importance of many of the historical rights is at least threefold: first, they gave the land flexibility and amenability of use for all; second, they provided a basis for the prevention of, or resistance to, hegemonic power over it; and last, the way that rights over land affected culture, identity, and attachment to land and environment would have been marked.

The particular outcome in terms of the way in which rights over land were distributed represents a rationalisation process, one that clinched control over the land (Donnelly, 1986; Thompson, 1993). This hegemonic assimilation of power was consolidated quite subtly in some ways. Hegemony can be viewed as a *process* rather than a *state*, much in the way that citizenship is conceptualised in this volume. Williams (1973: 112) argues that the maintenance of a status quo has 'continually to be renewed, re-created, defended and modified'. Conversely, this situation is 'continually resisted, limited, altered, and challenged' (*ibid.*) by those seeking to alter a status quo. Clarke and Critcher (1985: 228) view hegemony as a process 'involv[ing] the effort to dominate a society in which the divergent interests and perspectives always threaten to outrun the ability of the dominant culture to contain and incorporate them'. They identify leisure as an important facet of the struggle for hegemony in the UK in two ways: first, through the repression of 'undesirable' uses of free time, and second through the substitution of these with 'leisure patterns, which are civilising and profitable' (*ibid.*: 228). In the context of a contemporary countryside, increasingly valued for its amenity and leisure (consumption) use, this struggle is ever more important and increasingly places leisure activities as important points of both symbolic exchange and conflict.

This impacts directly on the way that citizenship might be constructed and the shape and form that land rights distributions should take. The 'loss' of local rights at one time and in one place represented rights that could have become (national) legal rights given different historical circumstances and rights that are today conceptualised as being national. Economic restructuring was consolidated and defended through rural institutions such as the Church. The attitude of the Church towards popular recreations such as rough football and quasi-pagan festivals was that they tended to run contrary to the accepted tenets of 'regularity, orderliness, sobriety, providence, and dutifulness' (Malcolmson, 1973: 90). It was argued that they encouraged moral laxity and as such were to be discouraged in order that 'individual and social discipline' could be observed. Tawney (1926) takes a contrary view, identifying the Church, during the earlier periods of change, in the Middle Ages, as one of the main opponents of such wholesale socio-economic change and the subsequent effects on rural society. The Church was often the protector of the poor, standing up for the interests of the peasantry. Charlesworth (1980: 105) reinforces this point:

> at the very moment when developments in agrarian capitalism should have torn down the veil of paternalism, the persistence of the gentry and the clergy in upholding their time honoured roles as guardians of the poor gave the needed legitimation to any defence the labourers might attempt of their traditional rights under that code.

Malcolmson (1973: 74) remarks, contra to Charlesworth and Tawney, that in England 'the established Church was largely a senior servant in the machinery of government'. This meant that in the long run the Church upheld the interests of social stability and the vested interests of the state and that the Church helped implicitly, if not explicitly, to make acceptable and reinforce the new order of things. It so happened that this 'new order' provided a convenient social shell for behaviour more fitting with a 'Christian lifestyle'. It also meant that, fortuitously, the Church's lands were to be more profitable and valuable in unfettered ownership. Chambers and Mingay (1966) point out that some clergy encouraged enclosure as it enhanced the rental value of Church land, or in some cases their own land. They were certainly not supportive of 'revolutionary' ideas that were under-pinned by early communism. A famous case involved the local parson and landowner Parson Platt ousting the Diggers at Cobham in 1650 (see Boulton, 1999; Bradstock, 1997; and Chapter 5). There is little doubt, however, that on a local level the apparent inequities caused by the changes in ownership, custom and lifestyle were, generally, opposed by the clergy (Hobsbawm and Rude, 1973). It is important to remember that continued resistance to social and economic change was difficult to maintain. Even now there are still fragments of old paternalistic/feudalistic attitudes prevailing in many areas of the English countryside, as there are examples of resistance and challenge to such power structures that echo struggles for rights since the time of the Diggers.

It is the case that land ownership and existing rights distributions are exalted. In 'advanced' Western societies they are reified and presented as natural. The present structure favours the interests of the power elite (see Lefebvre, 1991; Woods, 1997). If we look at the enclosures contemporan-eously, during a period of agricultural restructuring and commoditisation of the countryside, it is interesting to draw parallels between these old extinguished rights and the present moves towards the separation and monetisation of similar rights. Those, of course, now lie within the bundle of rights held under the now dominant legal construction of private land ownership. There is perhaps a case for 'new' rights that would be appropriate for the needs of a post-industrial society. Relearning land and land use and reworking the contract require an aspect of cultural and technical 'competence' and respect for land other than as commodity or production space. Bromley (1991, 1998) has argued that this might be impractical without better understanding of and more adept social and cultural 'ownership'. Society will need to relearn or reconnect with land as well as understand it as 'social space': as part of the processual dimension of citizenship. Calls such as Fairlie's (1996), for the restructuring of systems of governance to allow for a more sustainable and flexible approach to land, can be seen as a practical and useful step in reflecting economic and environmental priorities and social change. However, the notion that numbers of people will want to manage land in any direct sense is an unlikely one. Below, protest and the countryside is examined in its historical context, linking into the examples of conflict highlighted throughout the text.

Rural protest and the resistance to change

The countryside, particularly the English countryside, has been the backdrop and the prompt for very many protests over various aspects of agrarian life, including mechanisation, subsistence rights, access to land for recreation and, latterly, resistance to development for environmental motives (see, for example, Mingay, 1989, 1994; Thompson, 1993). Such agrarian-based protests have echoed down the centuries, with the forms and underlying reasons for those protests remaining remarkably familiar (see Thirsk, 1999). In recent years protest over land economy, use and access have all proliferated and the way they have been represented has intensified. This theme of conflict and contestation forms an important part of citizenship and of countryside culture.

Disputes over the rightful distribution of rights in the land have existed throughout history. In England the development and observance of many common rights were customary and therefore without the force of law. Indeed, customary rights were said by Sir Edward Coke, the distinguished judge, to be 'endless' (quoted in Thompson, 1993: 137), although attempts to disentangle the complexities of common rights and land have been made since (see DETR, 1998c; Gadsden, 1987; Hoskins, 1963). The landowning class were, in many cases, the creators and enforcers of the law: as legislature and judiciary, with both being heavily represented by landowners. This meant that the rights-claims of many commoners and landless people were lost during the process of enclosure, parliamentary or otherwise (Tawney,

1926; Norton-Taylor, 1982). Prior to the enclosures, much more land was open and the regime of governance was irregular, or uneven, based on custom and a culture of reciprocity. Of course, there was also far less regulation of land use or development. Shifts in the perceptual and actual role of land during the enclosures were met with furious resistance, but as the process was piecemeal, so was the protest: paradoxically, the modernisation of one area or 'social cage' (Mann, 1987) was made easier because there was a much weaker 'national' culture that might have rallied in a more unified way against enclosure. Developments in the modern era mean more or less comprehensive governance and also better-networked and better-informed populations. This is certainly a subject worthy of further research, in particular to assess local rights and local 'difference', and how such difference is sustained.

Various types and acts of protest are recorded, the motivations of which centred on a range of lost 'rights':

> In Wales and in some English districts, many of the crimes of theft, and some of violence, occurred on disputed or newly enclosed land. What caused particular annoyance was the legislation defining ownership of wild produce, birds, fish and animals.
>
> (Jones, 1989: 115)

The enclosures and the rationalisation of land use meant that the types of activities exercised through *de facto* rights were curtailed. The landowners enforced their right of exclusivity, and right to use the land as they wished, in order to carry out more intensive agricultural activity. 'Crimes of trespass formed a small but persistent element in the statistics of rural crime. . . . Similarly, prosecutions for trespass and for the destruction of weirs, walls, trees and produce could indicate battles over disputed property and rights of way' (*ibid.*: 119).

The demolition of fences and filling of ditches marked one form of physical protest. Other 'crimes' were linked to the changes in the socio-economic make-up; they also represented resistance to the difficult forging of modern citizens and citizenship (and notwithstanding the concomitant drive towards nation/state-building; see Dodgshon, 1999; Jones, 1999). These were part of the active resistance to the enclosures (Hobsbawm and Rude, 1973; Shoard, 1987; Tawney, 1926), but of course the picture was vastly more complex than this. Trade-offs were made around the country; some protests were successful or some landlords were so benevolent as to make decisions that prioritised the social over the economic (see Hill, 1996). Many people were attracted to the towns and cities to find better livings and improved conditions.

Historical discourse and the re-imagining of citizenship

New groupings, new sociations identified by Hetherington (1996) and Urry (1995) were beginning to identify and coalesce according to different priorities by the 1990s, notably in respect of the environmental agenda. A relevant example emerging in the mid-1990s is the land rights group The Land Is Ours (TLIO). This movement emerged to challenge both the traditionally accepted norms of citizenship and the boundaries of land and property rights in the UK. TLIO has linked citizenship to environmentalism and has in Anderson's terms (1997) become one of the new 'claims makers'. Its members challenge the decision-making processes affecting land and its management. Their objectives include access to land for amenity purposes (the first dimension of access), but they also aim to raise more fundamental issues concerning land. The TLIO aim is expressed as: 'The Land Is Ours campaigns peacefully for access to land, its resources and the decision-making processes affecting the land, for everyone, irrespective of race, age or gender' (TLIO, 1994, 1995).

The main objectives of The Land Is Ours campaign are to re-examine the way in which land rights are treated in their social context. TLIO poses moral questions concerning the use of derelict land, the issue of homelessness, the claim-rights of minorities such as travellers and ravers in relation to land, and confrontation with landowners' behaviour on a moral level. The campaign problematises rights not just of 'ownership' but also of empowerment: 'Our role is to highlight ordinary people's exclusion not only from the land itself but from the decision-making processes affecting it' (TLIO, 1995: 1). The movement follows closely in the footsteps of arguments already put forward: that part of the 'stake' of the citizen is a claim to rights in the land of the place from which their citizenship derives. Indeed, this resonates clearly now with the political rhetoric of the UK Labour Party, which also utilises the 'stake' and the 'contract' as important elements of the 'Third Way' (as discussed in Chapter 3). The new challenge is to reconcile old forms of land and governance with new conditions of social reflexivity and drastically altered global economic relationships.

TLIO symbolically encapsulates the main themes that are developed in this book. Present are the links between land, citizenship and the concurrent claims about land use in productive and consumptive terms. It was born of the emergent direct action movement that had been developing since the Greenham Common actions during the 1980s and into the roads protests of the early 1990s, notably with the Twyford Down protests in 1992 and the Newbury ones in 1995 (McKay, 1998; Halfacree, 1999; Merrick, 1996). The movement also incorporates concern about sustainability in social and environmental terms. TLIO can be seen as an example of a new sociation or community of interest that has as its focus land use and particularly, but not exclusively, rural land and land rights. One of its actions is used as a case study in Chapter 5.

Much as Gerrard Winstanley had over 350 years ago, TLIO have an alternative vision for rural land and its governance. At the time of writing the campaign had successfully drawn attention to land issues in the UK with high-profile occupations of St George's Hill in Surrey (the site of the original Diggers' commune) in 1995 and again in 1999 (see Chapter 5; Jefferies, 2000). Similarly, the 'irresponsible' ownership of land has been highlighted through exercises such as 'Pure Genius', where derelict land in London was occupied and an alternative community set up in order to emphasise the possibility of alternative land use and low-impact development on both rural and urban land (Halfacree, 1999; TLIO, 1995). By 2000 the movement had begun lobbying government and other agencies and interest groups on land issues as part of the wider rural/environment/planning policy community.

Citizenship is about the rights and responsibilities of the population; this includes the identification of citizens with land, but it is also about the contingent and uneasy relationship between the social and the state (see, for example, Mann, 1986). The TLIO campaign can be read in terms of postmodern citizenship. It represents a claim based not on class, but on consumption, the environment, on historical claims and lifestyle, including alternative (rural) lifestyles being pioneered by some of its members. It is also a positive campaign with a clear set of objectives rather than simply presenting reactionary opposition to existing distributions. It highlights the linkage between exclusion from the land in a physical and economic sense and also from structures of governance – a problem that the planning system, for example, has struggled to overcome in the past (Fairlie, 1996; Hall *et al.*, 1993). The following section picks up on two key words that have already been used liberally: *access* and *trespass*.

Trespass, middle-class 'resistance' and countryside amenity groups

Access is a curious word in the English language, one with several meanings. Colloquially, when applied to land it has two main connotations, both of which are invoked by TLIO. The first is that of access for amenity or informal recreation use. The second, a wider meaning, is that of access to land to use it productively or for shelter – that is, as a means of combating socio-economic exclusion. The debate over access (for amenity use) provides a single-issue approach by which to get to the heart of issues surrounding the citizenship and land rights debate, and in recent years the long-running debate over access has climaxed with a government commitment and parliamentary bill for a 'right to roam' (DETR, 1999).

In the nineteenth century the burgeoning population of both the north and the south of England became increasingly restricted temporally, spatially and financially (Stephenson, 1989; Mann, 1987; Douglas, 1976). Over time, improvements in pay and conditions enabled more people to take holidays and they therefore began to seek to make more use of the countryside as a recreational space. For others, unemployment and other aspects of urban deprivation meant that, paradoxically, they had no money but did have 'free' time to spare (Clarke and Critcher, 1985). The spectrum of circumstances that people experienced during the first fifty years of the twentieth century mainly conspired to increase the demand for access for amenity purposes. This set up a tension between the rural as Britain's 'breadbasket' – that is, the productive heart of the nation (and landowners as stewards of the land) – and the countryside as leisure space for the population at large: 'the nation's playground' (Williams, 1973; Shoard, 1999).

Claims for access were couched in terms of 'freedom', and one of the origins of rambling (working-class rambling at least) was to enable 'a retention of ties with rural origins' (Donnelly, 1986: 218). Such ties were most obviously linked to the recollection and imagination of prior relationships to the land: to rural space and, of course, land-based activity. Specific demands for access for informal recreation (i.e. walking) were linked to the old tramping tradition that involved workers from the countryside walking to work to neighbouring villages and in the towns, and to enable communication across country. Tramping between settlements had been taking place since prehistory. Indeed, many footpaths and rights of way owe their existence to these ancient usages (Malcolmson, 1973; Shoard, 1987).

This, combined with the 'romanticisation' of the countryside by notables such as John Ruskin and William Wordsworth, persuaded the city-dweller that the countryside was a safe, beautiful and healthy place in which to spend time. This notion countered previous, and intermittent, concerns that the countryside was 'a place of fear and dread . . . a gloomy and uncouth place' (Blunden and Curry, 1990: 21). These shifts in cultural perception towards the amenity value of the countryside led to calls for its protection and led the first countryside/amenity pressure groups to develop.

Issues concerning land, particularly agricultural prices and access to land in its wider sense, came to the fore around the time of the reforms of the 1830s (see Winter, 1996). This concern stemmed from both an urban and a rural perspective. By the 1830s unions were forming to represent the interests of the working classes. The Tolpuddle Martyrs case was heard in 1834, and by 1836 the People's Charter was published by a union demanding a range of reforms, including universal suffrage and a ban on property ownership conditions that barred entry for prospective MPs if they were landless. This charter formed the basis for the actions of the eponymous Chartists during the following years, which included demands for cheap land for the working class (Douglas, 1976; Trevelyan, 1967). Such objectives were progressed by the setting up of the Chartists' Co-operative Land Company to assist people to save and buy land. This was the period when the co-operative societies were forming (see Birchall, 1994). 'Productive' and private commercial use of rural land had, however, become more deeply accepted. It also tamed, beautified and disciplined rural space and rural people, as well as feeding a rapidly expanding national population. It is remarkable, however, to reflect that those defined as 'rural men' had to wait until 1884 for the franchise whereas their 'urban' brothers were eligible to vote by 1867 (Shoard, 1999), indicating how 'universal rights' were unattained by some and how citizenship was both contingent, multiple and exclusive.

Many of the countryside groups established in the nineteenth century were harking back to rural roots. As early as 1826, groups were forming with the purpose of protecting rights in the countryside. As noted, they were prompted by a variety of events: the loss of space brought about through enclosure; the romanticisation of the countryside by writers and artists of the period; and, latterly, the increase of leisure time, or at least the clearer delineation of work and leisure time being experienced by a wider cross-section of society (Clarke and Critcher, 1985; Blunden and Curry, 1990). This marked the beginning of a conflict over the use of land that is still current. The migration into the towns and cities left many people with a cultural void that was passed on to the following generations (Williams, 1973; Bunce, 1994).

By the mid-nineteenth century urbanisation had prompted the same modernising minds to begin to concentrate on the rudiments of city planning (see Cherry, 1973, 1984), but similarly, the perceived pressures on the countryside gave rise to a new preservationist ethic. The middle class invented 'planning'. City or urban planning and countryside planning arrived hand in hand, yet the former has retained its status as being high profile (reflected in the title of the Royal *Town* Planning Institute). It was definitely of, and for, modernity: helping to rationalise expansion and assist the project of modernisation. On the other hand, countryside planning has been cast in an oppositional role: forever to restrict the growth of the towns and cities. It has been seen as the resistant, now NIMBYist, partner to the 'progressive' yet monolithic Urban (see, for example, Cherry and Rogers, 1996). This binary division has had a number of ramifications since and has certainly been supported by historical and political circumstance, not least being both the world wars, which prioritised the agricultural and extensive agrarian use of the countryside. Such use has ensured that the ethic of stewardship and the need to 'protect' (rather than truly plan for) the countryside has remained strong until this day. Later it is argued that such a binary split is misplaced and that a more integrated and reflexive approach towards land-use planning should be developed.

Much resistance to change in the countryside in the nineteenth century stemmed from middle-class sources, and although this has not exclusively been the case, long-running protests and concern over urban sprawl and enclosure have filled English history. More generally the loss of green space prompted the rise of numerous amenity groups. The formation of groups such as the Commons and Open Spaces Preservation Society in 1865, the National Trust (1895), the Campaign for the Preservation of Rural England (1926), the Youth Hostels Association (1930) and the Ramblers'

Association (1935) came about as rural space became increasingly 'ordered', economically rational, monetised and exclusive.

The countryside was perceived as being increasingly distant to town-dwellers, and the groups aimed to protect and enhance its recreational and aesthetic value. The groups intended, in essence, to ensure that there was no further loss of green space and to try to re-establish certain rights to land. In particular,

> through the activities of the Commons Preservation Society, founded in 1865, the process [of enclosure] was checked, and the Commons Act of 1876 severely limits the rights to enclose ... although the Act of 1876 practically halted the enclosure movement, the destruction of the ancient manorial structure of villages had by then been almost completed.
>
> (Simpson, 1986: 261)

There were marked differences in the aims and approaches of the various groups, as Blunden and Curry (1990: 28) note:

> organisations such as the Commons, Open Spaces and Footpaths Preservation Society ... were more conservative organisations and lent more covert pressure for access. Membership of these organisations was largely middle class and they effectively used litigation, personal influence and connections with the Establishment to further their ends.

There were also quite recognisable differences in the groups' aims and make-up. It has been suggested that the Southern-based groups were largely middle class while in the North they were predominantly working class. This, it has been argued, determined their approaches and agendas into the twentieth century (Blunden and Curry, 1990).

The Arts and Crafts movement of the late nineteenth century wanted a working countryside as well as a beautiful and accessible one. William Morris and his *News from Nowhere* were closely related to the 'Back to the Land' movement of the late nineteenth century promulgated by other Victorian notables such as Joe Chamberlain (Douglas, 1976; Hall and Ward, 1998). This period represents a fascinatingly fluid time when the land and people's relation to it were in the balance. The modern social contract was very much in the crucible, with very many different issues and rights being claimed, apportioned and redistributed with direct and indirect impact on land and the relation of the people to it. Eventually the attempts to nationalise land and redistribute it were ended as much by war, the emergence of town planning and a subsidised agriculture as by any more sinister means. The period did, however, produce numerous acts of Parliament that reformed aspects of UK (and Irish) tenure arrangements, as well as the first compulsory purchase powers over land and the emergence of provisions for state-held smallholdings, allotments, and also the beginnings of the development of plotlands communities (see Hall and Ward, 1998; Douglas, 1976). In general terms, however, the period preceding the Great War did little to erode the power of private landlords, stem the tide of urbanisation or remediate alienation from the land of the majority. However, pressure did increase for land to be a *resource* for all.

After the First World War working-class amenity groups such as the disparate ramblers' groups (amalgamating in 1935 as the Ramblers' Association) were continuing a tradition of resistance to private and exclusive use of land. On a parallel course in terms of attempts to reform capitalist relations, the co-ops were thriving during this period. They provided working-class self-help in the face of a largely unfettered market system (Birchall, 1994). Various enterprises were set up to develop areas of rural land – generically known as plotlands even though they were cast in an unfavourable light by authors such as Thomas Sharp in his *Town and Country* (1932) as part of 'ribbon development' (the arms of Clough Williams-Ellis's 'octopus') and something to be vigorously discouraged. Many other groups were dissatisfied with the distribution and constitution of ownership, with rights and power over countryside goods, practices and working people living in the countryside (see Rothman, 1982; Donnelly, 1986; Newby, 1979, 1987).

The modern landscape did allow land to be utilised efficiently and, in times of national emergency, further exhortations towards efficiency and productivity were employed in order to feed and service the population more effectively (Wright, 1996; Tannahill, 1975; Newby, 1987). However, such productivity was seen to be at the expense of the urban majority, many of whom lived in dreadful conditions, and indeed working people in the countryside, many of whose living conditions were equally miserable. In this context, then, planning emerged as a modernist, technical and rationalist partner to the modernised countryside, in line with the aims of the garden city movement. For example, by the 1930s, plans for 'a hundred new towns' were advocated to reorganise the distribution of the population, plans that never came to fruition (see Cherry and Rogers, 1996) but which when part-implemented in the 1950s were to come to be widely regarded as a social failure.

The self-sufficiency priority in agricultural production (linked closely to the Victorian-puritan self-help creed; see Trevelyan, 1967) helped to safeguard rural land and its ownership by ensuring that agricultural land use retained primary importance. No credible challenges to rights distributions could be mounted or calls for land reform contemplated in the political environment of national insecurity (Winter, 1996). Cox (1984) and Cherry (1975) provide detailed histories of legislative struggles over the use and control of land in this period. They emphasise that the maintenance and reinforcement of a utilitarian conception of property rights, which best served the interests of 'the nation' while also serving the interests of landowners themselves, had been successfully accomplished. This trajectory was supported most discernibly in the powerful backing that the 1942 Scott Report, *Land Utilisation in Rural Areas* (MHLG, 1942), gave to self-sufficiency in agricultural production and therefore for agricultural land uses retaining primary status in the countryside. This underlying ethic is still discernible in agricultural/rural policy terms now (Cherry, 1975; Curry, 1993; Winter, 1996).

However, there has always been a strong minority view of a different vision of the countryside and for that matter of the relation between town and countryside. This was true even of the 1942 Scott committee, which was not unanimous about the way forward. The minority Dennison Report produced in contradiction to Scott shows how even during the Second World War a more diverse, and in retrospect essentially sustainable, countryside was being recommended, echoing the sentiments of the earlier back-to-the-land proponents (Cherry, 1975; Curry, 1993).

The post-Second World War years saw the continued, if not increased, protection of agriculture and therefore agricultural land and related interests. This has been evidenced in the strong corporatist influence that groups such as the National Farmers Union and the Country Landowners Association have had when policies involving land use in the countryside have been proposed. Interests of landowners and (larger) farmers were accorded priority and the constraining power of the agricultural land lobby remained largely intact into the 1980s (Winter, 1996). As discussed later, there are signs that the strength of the environmental lobby has begun to make significant inroads among this policy community. However, this is an increasingly complex and mediated policy environment, involving not only the agriculturalists but the environmental lobby, preservationist groups, animal rights interests, amenity groups and increasingly large numbers of the public concerned with safety issues generated by farming and other polluting activities in the countryside. All these are keen to voice their opinions and priorities. For government this presents a problem when attempting to be inclusive (or at least in appearing inclusive).

Trespass and protest

Land unsurprisingly lies at the heart of very many issues, causes and concerns. Land and access to it developed into a symbolic contest over rights and entitlements, at once a struggle for access and a struggle against perceived inequalities in society. It was appropriated as a symbolic form of protest generally, and specifically about social and economic exclusion as well as the functional aspect of public amenity use. This is one reason why this aspect of land and citizenship is used throughout the text to illustrate the overall themes. As previously shown, trespass, as an illegal or culturally inappropriate incursion onto land, has been and continues to be an important form of protest (Riddall and Trevelyan, 1992; Parker, 1997; Sibley, 1995). The mass trespasses of the 1930s were possibly the best-known access flashpoints. They are described here to illustrate how access has been used as an end in itself, but is now also part of a wider discourse aimed at 'reclaiming' the land, or at least aspects of the control of land. This is aimed at delivering a more equitable and perhaps flexible approach towards land use in the UK.

A preoccupation of the state has been to control such actions as part of attempts to maintain rights distributions, particularly in terms of private land rights. This is evident in recent public order legislation, for example the introduction of a criminal offence of aggravated trespass under the Criminal Justice and Public Order Act 1994 and other provisions restricting camping and other 'undesirable' uses of land (Card and Ward, 1996; Bucke and James, 1998; Parker, 1999a).

In much of the following section, and interwoven in subsequent chapters, we examine examples of trespass and reactions to such transgressions. Trespass can be viewed as both actual and symbolic transgression over lines of power, often plying a line between legitimacy and illegitimacy, therefore testing the state in terms of what may or may not be permissible. In this way, liminal or temporary autonomous zones are created, perhaps (see Shields, 1991; Bey, 1991). Similarly, rights of way and *de facto* paths can be seen as fissure points in relations of power and control over space. The challenge and counterchallenge represented by legislation and policy and legal decisions and public protest at least partly reflect the field of power and how different groups and interests stand in the relational sense. Thus, trespass has been used as a technique, by oppositional interests, to promote challenges to standing conditions (cf. the Diggers, mass trespasses, TLIO). Bey (1991) argues that transgression becomes a form of 'peak experience', one that retains meaning and power even though it is fleeting. Such acts of resistance may have a longer effect on power-holders, perhaps even, indirectly, in temporally changing policy in some contexts.

One expression of dissatisfaction came through the radical ramblers' groups, which saw maxims such as 'A Land Fit for Heroes' as empty and nationalistic tricks played by the state in order to galvanise working-class support in wartime. In the inter-war period tensions over poor social conditions were rising – especially in northern England (see Hill, 1980; Stephenson, 1989). The rambling groups, being steered by wider social and political ends (Donnelly, 1986), organised mass trespasses in the inter-war years as a means to publicise inequality using access to the countryside as the rallying point against other perceived inequalities. The first and most infamous mass trespass took place in April 1932 at Kinder Scout in the Peak District when around 400 people took part (Rothman, 1982). This incident served to show the frustration and feeling of inequity that existed between sections of the public and landowners, over the grievance that less privileged citizens felt at being excluded from large areas of open land. Open land was seen as suitable, accessible and affordable for recreation and it was felt that a 'moral right' to use the land existed – a claim legitimated by a particular reading of history (Rothman, 1982; cf. Ramblers' Association, 1998; Shoard, 1999).

The mass trespasses symbolised a struggle over rights in the countryside. They occurred at a time when the public were becoming more politically aware of the implications of private and public provision of services and the regulation of rights/rights-claims to those services. In one sense at least, some people were becoming more reflexive about the role of institutions in their lives and how this affected their citizenship 'envelope'. In terms of access to the land, they were concerned about the role of the state, the landowner and the market in controlling rights of all types. It was also a period when the institutionalisation and the nationalisation of services were gaining political currency (Cox, 1984). Given this context, landowners were generally more apprehensive and perhaps particularly anxious about ceding private rights, or making concessions, over land to the public. There were further trespasses later in 1932 and other rallies and demonstrations in support of improved access rights, some of which attracted five-figure attendances (see Lowerson, 1980; Rothman, 1982; Shoard, 1999). The largely working-class ramblers were 'forced into wars of attrition with landowners' (Donnelly, 1986: 219), rather than mounting expensive legal challenges over rights over land because of their economic and social circumstances. Those circumstances are reflected in reports of the trial of those charged after the first trespass, now known as the 'Kinder Scout Six'. They were unable to make a satisfactory defence for want of funds. The trial was rather unbalanced, if not prejudiced, with the entire jury comprising local landowners, local dignitaries and senior army officers (Rothman, 1982; Stephenson, 1989).

It is a strong thread, and one that has been continued by the Ramblers' Association in contemporary times, that many of the claims of the ramblers and political activists were to do with 'regaining' lost access rights. Such rights were claimed to have been lost during the eighteenth- and nineteenth-century parliamentary enclosures (Ramblers' Association, 1991; Donnelly, 1986; Simpson, 1986). As has been demonstrated, this may or may not have been the case, but what surely had been lost was the ability for land to be used flexibly and conjointly. The means and mechanisms for the democratic governance of land were yet to be developed such that interests other than those of the individual owners would be effectively and fairly considered. This issue is possibly still in need of further attention even now. One mechanism partly to remedy the issues being debated might be to carry through the plans of the post-war Labour government to nationalise land and develop a more expansive, flexible land-use planning system (Lichfield, 1965). The intensely political nature of the challenges that were made in 1932 threatened, or was perceived to threaten, the status quo and consequentially the dominant ideologies of property (Newby *et al.*, 1978).

Many attempts to challenge land ownership through countryside access failed, from the earliest in 1884 until the National Parks and Access to the

Countryside Act was passed in 1949. But this failed to deliver many of the desired freedoms (see Shoard, 1999; Parker, 1997; Parker and Ravenscroft, 1999; Blunden and Curry, 1990). One of the early parliamentary challenges, the 1932 Access Bill, failed to become law possibly as a backlash reaction to the mass trespasses. Pointedly, one of the speakers against the bill had decried it as 'a vicious and Bolshevik attack on private property rights' (Lowerson, 1980: 277). World politics and economics provided good reasons for landed interests to react so strongly. This was the era of the Great Depression, and the Russian Revolution which took place just fifteen years earlier, was threatening to internationalise communism.

Conclusion: land, conflict and citizenship definition

This chapter has sought to provide the contemporary discussion of land use and citizenship with an historical dimension that underlines how, for example, 'stake', contract and the development of the nation-state and citizenship have gone hand in hand with the regulation of the land and the construction of rights of private property. The historical perspective helps to illuminate the contingent and politically motivated underpinnings of rights structures in the UK, and England in particular. It illustrates how such an operationalisation of governance has been subject to continual, if only periodically visible, resistance and contestation and a rigorous, some-times brutal, defence.

The issues discussed are rooted in deeply political, if modernist, notions of equity, justice, liberty and equality. There are those who would wish to maintain, extend and reinforce many existing distributions of rights and responsibilities and those who would rather see some, if not all, of those outcomes and tenets of ownership altered. Added to this structural debate are the multiple, recent rural 'crises' that have developed over the course of the 1990s in terms of agricultural prices, environmental protection, food safety, deprivation and social exclusion issues, to name but a few (see, for example, Barnett and Scruton, 1999; Thirsk, 1999).

Through history there has been a struggle over who controls land use. One key means of resistance and protest has been to occupy or trespass on land. Periodically the state has promised reform and made an explicit connection between nation, people and the land. Such linkages serve as a reminder that the social contract involves a negotiated settlement about the 'best' form of arrangements for the governance and use of land. Fluid space and fluid society need fluid, responsive governance. In the era of the modern state, hegemonic manoeuvring has taken place, resulting in frequent concessions and trades being made; but certain distributions of rights and obligations become stuck, resisting further change. Important examples include the nationalisation of development rights through the planning system, and through policies such as the creation of national parks. Both can be (and have been) criticised as being partial and largely ineffective 'levellers'

(Ambrose, 1986; Shoard, 1987; Parker and Ravenscroft, 1999). Rather the state has reformulated the social contract primarily at times of national crisis. It is contended that the state is facing new crises both domestically and globally, the latter representing a threat to the state itself. This provides an opportunity to challenge historic assumptions, relations and definitions relating to citizenship and land.

One of the outcomes of the historical construction of the countryside has been the development of an imagined rural, bringing with it a sense of attachment and value to land and landscape and particular activities (e.g. leisure uses, and evidenced by tourism). This is coupled with an appreciation of particular built and cultural forms. Through representation these become idiomatic, almost hegemonic, forming part of the 'countryside aesthetic' (Harrison, 1991). This also impacts on citizen activity and identity. Similarly, counterculture drives the alternative agenda for land and regulation. These are explored through the research examples located in the following chapters where, for example, policy exhorting members of the public to be active is reviewed, and protest and the way that consumption impacts on the rural are used to situate the theoretical framework.

In the following chapters it is argued that citizenship should be viewed relationally, in terms of culture and in terms of contingency. Citizenship in the post-rural is both appropriated and resisted at all levels, and historical referents are used by those seeking to maintain the status quo as well as those seeking to effect radical change in terms of land and governance. As mentioned in Chapter 2, the synthesising relationship between people, land and artefacts and subsequent control within space has been labelled territoriality. Shifts in citizenship construction may therefore imply alterations to practice and the way that territoriality is effected and possibly eroded. The ways in which citizenship and power relations are altered include presentations of different readings of history. The discourse of stewardship deployed effectively in the past by dominant and landed interests is under strong challenge by groups claiming that the land should more literally be 'ours'. History shows how citizenship has been constructed and is contingent.

There are numerous justifications for and against various policy options: conflicting interpretations of history, of economic need, of social relevance and of political acceptability. The prescription of remedy in this context is fraught with danger. Each remedy will be criticised and numerous detrimental impacts are likely to be predicted. It is the search for alternative forms of legitimation that leads to the further, applied analyses of citizenship rights and responsibilities outlined in different contexts in the following chapters. It is also noted how certain forms of active citizenship can be read as attempts to mobilise discourse, drawing on the local and the global, in order to alter relations of power and to 'perform the state' (Albrow, 1996).

The contemporary use of the historical and the contemporary land use and planning agenda are looked at in the following chapters through the actions and rhetoric of the land rights campaign The Land Is Ours (TLIO).

All the following three chapters provide parallel examples of forms and expressions of citizenship in relation to the countryside. Specifically, Chapter 5 outlines two examples of conflict and rights-claims in contemporary politics. The narratives focus on the use of history to legitimise claims and actions in the present, linking back to issues raised in Chapter 2 and drawing on the political dimension set out in Chapter 3.

5 Enacting and contesting rights through history

Introduction: political action and citizenship

Chapter 4 provides an historical context for citizenship action and the construction of the citizen 'envelope' in relation to the countryside, particularly in terms of rural England and rural land. Chapters 2 and 3 also illustrate how citizenship is a contingent, politically constructed definition, but show that it can be viewed more widely. In Chapter 2 it was established that citizenship is widely considered to be more than the passive receipt of rights and responsibilities; it is also the creation, deliberation and contestation of such rights and responsibilities. It involves activity as well as status and is part of multilayered and multiscalar identity. Until later in the discussion the lack of traditional engagement on the part of the passive majority of citizens, in spite of feelings of democratic deficit, is put to one side. Rather, examination of aspects of non-traditionally accepted modes of citizenship is undertaken, particularly in terms of citizenship as 'protest'.

This chapter consolidates the link between conflict, protest, participation and citizenship with history and the land. Two case-study examples are used to highlight the way that agents attempt to alter rights distributions using a variety of means and tactics. The twin themes of history and resistance in the contemporary context are examined, showing how citizens participate in activities that generate change in terms of cultural and legal regulation. The way that citizens engage with others in political manoeuvre is assessed as a central part of citizenship as activity and as process, in particular by reopening 'black boxes' of stabilised relations represented by rights and responsibility distributions set in both legal and customary frameworks. One interesting approach to this strategy that is explored in the chapter is the invocation and use of history and heritage as a political tool. It is concluded that such efforts to challenge rights distributions may be detected in many policy areas and that increasingly mediated politics encourages the use of a heritage discourse and 'competitive storytelling' (Grant, 1994; Throgmorton, 1992) to destabilise dominant relations of power and open up possibilities for new claims to be authorised.

It is argued here that resistance can be seen as an attribute of citizenship, one that is perhaps as important as, if not more so than, compliance or state-defined 'good' behaviour. While Hegel asserted that participation was the highest of all human needs (Sabine and Thorson, 1973: 593), Oscar Wilde wrote that 'it is through disobedience that progress is made'. It is argued here that resistance (and constitutive forms of protest) can be viewed as both healthy aspects of citizenship and useful forms of participation in social and political life. Participation is something that should not necessarily be defined by one section of the polity only, or indeed defined in isolation by the state or by national government. Rather, public participation is expressed in numerous forms and legitimised, ignored or defined as transgressive by the state.

Citizenship involves the relations between the individual, society and the state, and the relations of these with the environment. Part of the contract that is being continually brokered involves the unsteady reconciliation of individual or minority views, desires and actions. Indeed, therein lies part of the trade-off between equality and individual freedom. The social contract is, however, always behind culture and wider events. In that sense the social contract, as expression of social relations, is imperfect temporally, and uneven spatially and in terms of diverse groups. It is being constantly amended and performed differentially. At one extreme, assimilation (as opposed to integration) of the individual into society involves the (enforced) adoption of existing rights, responsibilities and behaviour by that individual. Bourdieu (1977) sees this process as part of pedagogic action that may involve various forms of symbolic violence. The integration of these interests involves a much broader interpretation and requires a less rigid application of rights and responsibilities. This allows more scope for individualistic behaviour within the citizenship framework, in terms of both official sanction and imaginative engagement with authority.

Citizenship, destabilisation and dissent

An extensive 'pool' of alternative stories and interest claims are being articulated and contested over (rural) space (Murdoch, 1997a; Law, 1994). High-profile protest such as anti-nuclear protests, animal rights protests, industrial action and anti-road-building actions (to name but a few) have received much media attention. They have also been seen to 'make a difference' in policy, legal or cultural terms. Margaret Thatcher had wanted to eliminate such attempts to perform the state; for her, such action was not legitimate even though many other politicians and theorists accept that various sorts of protest are healthy forms of political action. Attempts to control such acts of resistance towards elected government or other unaccountable targets were, however, ultimately doomed to failure, and arguably the handling of the 'poll tax' 'riots' in London in 1990 was one of the seeds of Thatcher's downfall as Prime Minister. Effective control of protest is achieved largely by bounding and enrolling resistance, rather than by attempting to eradicate it entirely. Notwithstanding this, protest can be conceptualised as 'active' and, subjectively, as 'good' citizenship. On many occasions it is only in retrospect that such actions are deemed legitimate. It is also the case that sometimes the state is not seen to be 'reasonable', and media-ted protest achieves significant public opinion gains on two counts: both for the cause itself and on the liberty principle (Parker, 1999a).

Such high-profile actions are important, but it is also more small-scale and individual acts that form part of the process of remaking citizenship on the part of interest groups, the media and individuals. Indeed, such actions build social and human capital or help to capacity-build, both of which require knowledge, information and/or experience on the part of the 'citizen' (see Countryside Agency, 1999a; Healey, 1997; New Economics Foundation, 1998). This may not necessarily, however, be on the terms of the state and therefore is not often within the bounds of the arbitrary citizenship enve-lope constructed by the state. The point is that such behaviour or practice is variously defined as illicit, undesirable, immoral, deviant or simply illegal. In similar fashion, protest is viewed as threatening to the integrity and the stability of the state, and as such it is not considered to be part of good or even active citizenship. Rather, good citizenship is supposed to involve, and is portrayed as involving, acts within the envelope defined by the state; alter-natively, 'acceptable' citizenship is that which is defined in the cultural field (Bourdieu and Passeron, 1977).

The role of the citizen was thus to operate within legal frameworks handed down from government; amendments to such regulatory frames were to be made only through due process such as case law and other authoritative statements. While the citizenship envelope is constructed politically, and hence the delimitations of citizenship and the stipulations concerning entry as citizen are constructed politically, it is the case that such constructions are mediated by culture, and increasingly through what Fukuyama terms 'mediating institutions' (1992: 322). Foucault (1977) outlines importantly that deviance implies normality in a reciprocal relationship: deviant behav-iour defines what is normal. Bauman (1992) also describes the way in which the very existence of something that is portrayed as 'wrong' implies and helps define what is considered to be 'right'. Notwithstanding this, it is argued here that both protest and participation are modes of expressing the contestation of definitions and priorities; and are both political and cultural. Following the argument that citizenship is processual, in line with the contingency of rights (and responsibilities), it is the case that such arrange-ments are subject to perpetual challenge from various parties using various discursive means to influence change.

It is the impact of culture both at the level of the locality and at the level of community (of interest) that is of particular concern here, along with the tactics employed by citizens in order to claim rights and how this is done legitimately and persuasively. Before I outline the case studies it is useful to relate expressly the conceptual markers that are applied in order to explain and understand the strategies and outcomes of (micro)political agency. As mentioned in Chapter 3, manoeuvres or 'performances' (Albrow, 1996) shape state/legal definitions and distributions. This is perhaps particularly so when considering the wider community of interest in rural affairs and the contested category of 'rural space'. Earlier, territoriality was mentioned in relation to the implicit spatial dimension. Sack (1986: 19) links people, place and politics, outlining how rights/responsibilities may be shaped and contested in the process of maintaining or changing territoriality, where 'an individual or group [attempts to] affect, influence, or control people, phenomena, and relationships, by delimiting and asserting control over a geographical area'. In a spatial context both Sibley (1992) and Cresswell (1996) have examined the way in which space is thus 'purified' politically and culturally in order to allow practices of dominant groups to take place at the expense of other, marginalised activities. Cresswell (1996) comes close to a Bourdieuian analysis when underlining that the normalisation of certain practices, along with a range of other consolidatory factors, means that people's place in space can be categorised as either legitimate or illegitimate. Murdoch (1997a) quotes Leigh Star in identifying 'space as an ordering of priorities': while priorities may be individual, they are also subject to a range of strictures, and in terms of space/time certain priorities are enforced and do not allow for others to be present or visible.

Such questions about what practices are considered to be appropriate are complex; historical referents, for example, can rupture contemporary alignments of power. Assessing them requires the consideration of localised power relations, local practice and the influence of extra-local influences. In this respect Bourdieu's work on distinction and his concepts of habitus and field are useful tools in conceptualising local social/political relations (see Bourdieu, 1977; Jenkins, 1992; Fine, 2001). Bourdieu (1977, 1984) argues that on the level of the (cultural) field, norms and conventions are born of routinisation, pedagogic action and the internalisation of extant power relations. Bijker and Law (1992: 9) see that they are 'born out of conflict, difference and resistance', while Clegg (1989) sees stratagems or 'plays' as exercises of power made to determine whose definition should prevail. It may also be argued that where there are attempts to alter right/responsibility distributions, such occurrences, if successful, can be seen as 'moments of translation' – that is, when an alternative definition becomes more dominant, accepted or given authoritative status. Callon (1991) states that a successful process of translation 'generates a shared space, equivalence and commensurability', which suggests that a given issue or potential argument over a rights distribution should become relatively stable for a critical period. However, it is unlikely that resistance and counter-plays are rendered impossible. In Bourdieuian terms such plays and the resistance to them are expressed as part of symbolic violence (see Parker 1999b).

Modernist thinking can be critiqued (in much the same way that Marshallian citizenship theory has been criticised for seeing rights as unidirectional and evolutionary; see Chapter 2). Fukuyama (1992) may contest the evolutionary, but here such directional thinking about history is also contested. History and time are resources deployed unevenly and strategically in directing trajectories of change and continuity. It is argued that 'cultural competence' is not fixed in the present, but can be derived or imagineered from the past and be fluid and fragmented, thus bringing into question issues about how and by whom such competences are assessed and defined. As part of the processes outlined in Chapter 4 in terms of history and historicity, Bender (1993: 260) argues that historical and archaeological precedents were used to formalise 'customs and tradition into instruments of government and a defined code of laws'. Such projects were aimed at stabilisation, regularisation and homogenisation. Practice and culture, however, are inherently unstable or liquid. The construction of space is dependent on distributions of power and plays are made to alter extant power relations. Such alterations may be gradual and organic, or be more planned and noticeably dissonant with the previous situation(s). The examples given below relate to two aspects of the above. The first is how a 'protest group' found a moment of praxis or where 'Third Space' found practical expression (Routledge, 1997; Soja, 1996). Alternative practice in terms of land use was articulated. The second is where an individual through small acts gradually came to challenge others over a policy issue and rights/responsibility conventions. Both are portrayed as being representative of alternative, perhaps postmodern, forms of active or engaged citizenship. As examined in this chapter and then in Chapter 6, attempts made in different ways to alter such relationships of power are set out in order to highlight such flows (see Appadurai, 1990).

Citizenship as manipulating space and time

Game (1991: 26) notes that Walter Benjamin argued for a 'blasting open of the continuum of history' and for brushing 'history against the grain'. Benjamin's statement is read here as being a call to challenge notions of unidirectionality or 'progress' itself. Culture can lead policy in the sense that gradual change and fluid relations demand change on the part of structuring authority. This underlines how citizenship as reflection of culture is also processual, rather than simply a defined and staccato-like or laggardly legal status. Such processes also influence the way that land and space are used and imagined. Bauman (2000: 110) notes that modernity is centrally concerned with fluidity or liquidity and that time under modernity is 'the time when time has a history'.

On the level of institutions, fluid relations are more likely to demand more flexible institutions and other key actors such as landowners (perhaps more 'brandholders' in the semiotic, consumption-oriented sense; see Chapter 7) to become more aware of legal and cultural rights and responsibilities (Massey, 2000; Bromley, 1998; Blomley, 1994), as well as seeking to commodify benefits held as rights. Citizenship is partly about the rights and responsibilities of the population and the actions and identification of citizens with land, but it is also about the contingent and uneasy relationship between the social and the state (see, for example, Mann, 1987; Turner, 1986). More single-issue groups, direct action movements and non-governmental organisations (NGOs) such as Greenpeace and Friends of the Earth, alongside other associations based on communities of interest, have formed and have engaged with other powerful actors involved in glocalised governance.

Another outcome of changing social relations and reflexivity has been the rise of DIY culture and lifestyle politics, and associated increasing levels of public awareness of issues that affect the 'everyday', the 'routine' or the 'conjunctural' (Shotter, 1993; Gramsci, 1971). Such reflexivity is reflected in decision-making and in modes of political participation and protest rather than in terms of scale or extent of formal participation in political systems – for example, voting or attending public meetings. In that formal sense, 'civic sclerosis', as mentioned (see Alinski, 1972; Selman, 1996), appears to be increasing, or at least not diminishing (see Johnston *et al.*, 2000). Alternative modes of engagement are developing through technology and consumption practices (Parker, 1999c; Urry, 1995, 2000; Klein, 2000). This has been increasingly occurring, therefore, outside of government-delimited notions of 'active' or 'good' citizenship, and is effected by using different tactics and intermediaries, for example non-violent direct action or media manipulation.

On the level of agency, the use of aspects of the sociology of translation is useful here (see Callon, 1986, 1991; Law and Hassard, 1999; Murdoch, 1997b). It incorporates discussion of how networks pleat and fold time–space through the mobilisations, cumulations and recombinations that link subjects, objects, domains and locales (see also Serres and Latour, 1995). It is argued here that it is actors who engineer such pleats and folds and it is power that enables such pleats and folds to be made (see Dugdale, 1999). Such actions match and contest times, places and practice with power relations and preferred (political) futures. Actors may call upon distant (and temporally distant) resources to justify present and future action and therefore to (attempt to) determine futures. In this way proponents put forward alternative claims to legitimacy as part of a reflexive post-historicist project.

Material resources such as texts may preserve social order, power, scale and even hierarchy. Texts, however, may or may not suit a network in its contemporary position and may be used, ignored or undermined by actors; or alternative texts (or simply alternative readings of those texts) can be used to undermine or subvert dominant readings. In this way the manoeuvring of texts can perhaps create new possibilities and, particularly when considering culture and semiotics, do service for particular (political) projects. Murdoch (1997b) refers to Mol and Law's (1994: 663) concept of space as being 'a question of the network elements and the way they hang together. Places with a similar set of elements and similar relations between them are close to one another, and those with different elements or relations are far apart'. This helps to enable us to think of farness and nearness as being involved in the process of retrieval and actualisation of such resources by actors in their attempts to make stronger, closer and more stable relationships between places, times and practices. Although this is not necessarily more conventional, they may be attempting to rework the cultural arbitrary through such 'strategic games' (see Simons, 1995 on Foucault). 'Distant' resources or texts may be particularly useful to actors by virtue of their opacity or reification over time. In this way, reflexivity and historicity can be used to plan the pleating and folding of space and time; this disorders and ruptures trajectories and makes them contingent and non-linear. Both Wright (1983) and more recently Jefferies (2000) note how George Orwell foresaw the control or manipulation of history (including perhaps 'heritage') as important in influencing the present and the future.

Both Bauman (1998) and Beck (1998) have recently discussed how the present is discounted in favour of future thinking. Considerations for the 'future' are increasing preoccupations, but our imaginings are more and more weighed down by the past and by the burden of recorded and semi-imagined history. Giddens regards such reflexivity as being implicated in historicity – where agents use history 'in order to change history' (Giddens, 1984: 374). Hewison (1987), in bemoaning the erosive qualities of museumification, asserts that 'as the past grows around us, creative energies are lost. Worse, as the past receives more attention, it becomes more attractive, and the present correspondingly less so.' The idea that the past 'grows' is interesting, and yet the notion does not appear to allow for creative reflexivity. The disruption and reinvention of the present and trajectories of the future can be manipulated by representing alternative histories. In that sense, portions of time and practice that are (potential) alternative and present alternative futures are latent discursive resources. It is also arguable that such a process nullifies other processes of change and innovation. Rather, the 'saving' of fragments of time is done in the full knowledge that change and innovation imply a need for a reciprocal conservation of the outmoded or obsolescent. In one sense at least, Hewison's (1987) observation that museums have proliferated rapidly since the 1970s can be extended and contrasted to a similar 'museumification' of space through national parks, conservation areas, national nature reserves and a range of other preservationist/amenity designations in and of the countryside. Culture cannot be frozen so easily, and consideration of the impacts of associated 'heritagisation' of the countryside is an important project (see Agyeman and Kinsman, 1997).

Of particular interest here is how such actions and stratagems constitute attempts to rework citizenship and associated rights distributions, and also how such activity of itself constitutes examples of active postmodern and reflexive citizenship. Here, specific examples where actors have engaged in the practice of what Routledge (1997) terms 'aestheticised, postmodern politics' are examined. These are narrated to examine how spatial practices from the past and the present are being used as part of attempts to challenge (*qua* active citizenship) dominant constitutions of societal constructs of space and practice.

Digging and invading: history and the 'reservoir' of time

The historical context detailed in Chapter 4 provides a context for this section. Brief examples were set out to indicate certain key moments in rural politics. The historical thread running through this work is taken up again here, showing how past/passed time can be retrieved and mobilised. The process of imagineering the future using the past to contest politics of the present involves the representation of place and practice. One form of this is to influence power relations through the representation of history in order to contest trajectories of rights and responsibilities for the future. Part of such representation calls upon old practices to destabilise or at least question present activity and establish different practices.

Representations of time and history (as 'heritage') may, as explored in what follows, involve space and practice and attempts to align networks by presenting preferred future constitutions of society through particular representations of the past. This is congruent to the analytical proposition of Short (1991: 5), who observes that myths can 'destroy time'. Thus attempts to pursue rights-claims and to influence culture in this way, with a view to moulding practice and rights and responsibilities, may be integral to interests looking to advance their cause. Fragments of history hold different cultural or semiotic power. Bender (1993: 275) notes this, and how different conceptions of the 'preferred' jostle for position and are deployed out of time and in some sense 'out of place': 'a cacophony of voices and landscapes through time, mobilising different histories, differentially empowered, fragmented perhaps, but explicable within the historical particularity of British social relations and a larger global economy'.

Employing alternative discourses in the course of such folding and pleating to bring new resources to bear on an issue may disrupt 'normality' or dominant views of space and practice. Culturally, heritage is viewed as virtuous or at least benign – it holds symbolic power. Such power can be tapped or mobilised by actors in order to further their own interests or the interests of a network/community. Such mobilisations, however, require the selective use of information and historical knowledges. Historical representation is effected using parcels of time *qua* heritage as reservoirs of latent, preserved or 'salted-down' power. In essence this accords with the metaphor that Latour (1994) uses in referring to texts as representing the 'congealed labour of absent others'. Through texts and other artefacts (including the landscape and other features of place/space), preferred indicators of past nature–society relations are drawn upon (see Macnaghten and Urry, 1998; Tindale, 1998; Bender, 1993).

Below it is shown, through two examples, how groups and individuals seek to mobilise parcels of 'significant' time and alternative practice in order to progress and consolidate a project. When applied to examples of postmodern politics, aspects of translation theory can be usefully employed to help explain the process of mobilising, organising and reproducing resources in this way (see Dugdale, 1999). In both cases discussed below, time and history are reworked and manoeuvred using what can be labelled as a 'heritage discourse', whereby actors exploit imagined place/practice as part of their attempts to rework citizenship. The examples relate to disputing ownership, rights and management, but involve the spectacularisation of historic and symbolic spaces as 'action spaces' (Goffman, 1967). This may involve the creation of zones of temporary autonomy (Bey, 1991; Routledge, 1997) in order to attain 'critical dominance' (Jessop, 1990) so that the claim can be realised.

The Diggers: rights and remembrance

As part of an increasingly reflexive society and a linked awareness of historicity, radical groups and others (perhaps definable as active citizens) look to actions and spaces that have been lent cultural authority by 'history' and where cultural distinction is afforded by transforming history into 'heritage' (see, for example, Hewison, 1987; Lowenthal, 1985). Such validation is contestable and subject to revision, but history as past practice and 'past place' can be mobilised discursively and actively as part of contemporary projects. This example serves therefore as an exploration of how a discourse of heritage has been deployed rhetorically and spatially as a method of progressing a radical political project.

This first of two examples relates directly to the TLIO actions that were effected during the mid to late 1990s as part of a campaign to raise the profile of land rights issues in the UK. The background to TLIO was mentioned, in summary form, in Chapter 4 (see also Halfacree, 1999; Jefferies, 2000). A subgroup of TLIO (although not linked to it officially), known as Diggers350, orchestrated an example where history was made to do service for rights-claims in the present. The focus of the narrative concerns a series of linked events organised by Diggers350 in the spring of 1999 including a conference and an occupation of St George's Hill, Surrey, England. The exercise was given further resonance by the fact that St George's Hill has been developed into a highly exclusive private estate and golf course. The primary motive for selecting the hill was that in 1649 it had been the setting for an occupation where the occupiers aimed to settle and make productive use of the common or 'waste' land (see Chapter 4).

A band known as the 'Diggers' or 'True Levellers' occupied the site then. From April 1649 the group established a commune on the hill until their eviction and relocation to nearby Cobham in August of that year, whereupon they continued to practise 'Digging' as they had done at St George's Hill. Their actions represented direct challenges to the emerging Cromwellian regime over its position on land rights. Their aims were more radical than those of the Levellers, led by figures such as John Lilburne (see Hill, 1996), as they aimed eventually to bring all English land into 'common ownership'. This was to be part of a social revolution to reinforce the political revolution taking place after the Civil War. The Diggers' approach spread across England in the early 1650s, causing significant political controversy. The army crushed the social project envisaged by their best-known leader, Gerrard Winstanley, even though it did serve to reignite tensions over enclosure and the rights of (Norman) landlords over (Saxon) land (see Bradstock, 1997; Trevelyan, 1967; Tawney, 1926).

Prior to the 1999 action TLIO had staged its first occupation, of a site adjacent to the hill, in 1995. This event brought its own publicity and sparked off a fresh debate about land rights (TLIO, 1995). It also prompted several of the activists within TLIO to further research the Diggers, Winstanley and St George's Hill in more depth (Lodge, 1999). This process of learning by a small group led to the planning of the events of 1999 to mark the 350th anniversary of the Diggers' communal experiment. In line with TLIO thinking, the action was to have a clear set of objectives, rather than simply presenting reactionary opposition to existing distributions or policies.

A public conference, the Hearts and Spades conference, was held in nearby Walton the following weekend and was addressed by academics and local historians (see Bradstock, 2001). This event further raised the profile of the Diggers and in particular of Gerrard Winstanley. This conference acted as a means of providing prestige and symbolic capital for the Diggers both old and new. By association, it gave a degree of legitimation for the ongoing action and the events of 1995, which aimed to raise the profile of the Diggers and of Winstanley as a political figure. One of the reasons that Winstanley in particular has become a focus is the writings that he produced, which have survived and been amplified by numerous historians. His texts now represent resources that Diggers350 draw upon to legitimate their own activities. For example, much use is made of quotations from his pamphlets by the group and more widely by TLIO.

The 1999 action was undertaken in order to publicise the Diggers350 themselves, the original 1649 Diggers, and the ideas that both shared for a fairer distribution of land. The occupation was also in line with the wider agenda for land reform and approach being pursued by TLIO. St George's Hill was occupied on 3 April by around 300 marchers, who brought with them a pre-prepared memorial stone (Lodge, 1999; Bebbington, 1999). The marchers gathered, and readings of Winstanley's works were made; in a truly postmodern twist, the actor who played Winstanley in the eponymous 1970s film directed by Kevin Brownlow was enrolled to read some of Winstanley's prose.

Soon after arriving at the hill, a smaller group established a camp on the site (TLIO, 1999). During the occupation the memorial stone (Plate 5.1) was installed and the occupiers dug and planted a variety of food crops – just as the original group had reportedly done (see Bradstock, 1997; Petegorsky, 1995; Sabine, 1965; Tawney, 1926). In this way the past practices of the Diggers were enacted or performed. The action can also be seen as an attempt to create a temporary autonomous zone (TAZ) or 'action space' (Bey, 1991; Goffman, 1967); it was both symbolic and 'lived' in the sense that people who stayed on-site adopted historical practices until they were evicted. The action continued until 15 April when an eviction order was issued by the high court in London. The occupiers then moved off the site peaceably, knowing that the main objectives of commemoration and press coverage had been achieved (TLIO, 1999).

The occupation of 1999 generated much interest and controversy in the locality (*Surrey Comet*, 1999, 2000; *Guardian*, 1999). There was both opposition and support. The owners of the golf course site made it clear that they would not want the memorial stone to be left on the hill permanently. Their main objection was that people would trespass in order to view the memorial. Other local people, however, were supportive and, as mentioned below, support for the memorialisation (as opposed to the longer-term objectives of Diggers350 and TLIO) was expressed.

The authority of history, congealed partly through Winstanley's texts and by subsequent historians, that was tapped by Diggers350 appeared to play a significant part in convincing elements of the local community that their actions were not 'mindless' or without purpose. This aspect of the action is evidenced by the second, and ongoing, part of the narrative of this example. This involves the mobilisation of parts of the community around St George's Hill in actively participating in permanently memorialising the Diggers.

Commemoration and memorialisation

This part of the narrative describes a process of enrolling local support for a funding bid to memorialise the Diggers in order to educate people about them and the places they were associated with. The bid taps into government policy in terms of local community involvement in heritage planning and management. One of the main vehicles designed to do this is the Local Heritage Initiative (LHI) run by the Countryside Agency and through which the Diggers' bid has been made (Countryside Agency, 2000a). The LHI project to memorialise the Diggers involves the installation of the original memorial stone that had been made for the 1999 event and the installation of other memorials and interpretation boards at numerous local sites with connections to the Diggers (Taylor, 2000). The plan is for a branded 'Diggers Walk' to be 'created' in and around Weybridge and Cobham in Surrey to connect the sites (Countryside Agency 2000b; *Surrey Comet*, 2000).

Plate 5.1 The Gerrard Winstanley memorial stone *in situ* at Weybridge, Surrey. (*Photo*: Gavin Parker)

The local media, which followed the 1995 and 1999 events with a degree of hostility (*Surrey Comet*, 1999), have reported on the project, greeting it quite positively (*Surrey Comet*, 2000). The project has also attracted attention from the BBC, which screened a piece on the Diggers in September 2000. The local network has drawn upon a national-level initiative in order to further aspects of the Diggers350 project. In Chapter 6 a similar process can be seen in the operation of the Parish Paths Partnership scheme, another national scheme for aspects of rural land management (Countryside Commission, 1994; Parker, 1999a), in which some factions of local populations used the scheme to destabilise the cultural arbitrary.

The chairman of the Countryside Agency, Ewen Cameron, underlines the rationale for the LHI scheme: 'the beauty of the Local Heritage Initiative

is that it allows communities themselves to decide what is important'
(quoted in Countryside Agency, 2000c: 1). The LHI therefore allows
groups to put forward a case at the national level that their application for
funding (and therefore authentication) is, first, 'heritage' and, second, is
appropriate to have public money spent on it. It is also assumed that such
projects are not in conflict with other views or conceptualisations of heritage.
LHI has the potential to privilege certain aspects of (rural) culture and
history, effectively silencing or marginalising other rural imaginations. It
must also be said that TLIO and the Diggers350 organisers have risked
depoliticising the Diggers' message by 'heritagisation'. However, a counter-
argument to this is that the Diggers' message can be used elsewhere and at
other times – in the future TLIO and other groups may still appropriate the
Hill for their political purposes. Indeed, it can be argued that the events at
the Hill in 1999 inspired the 'guerrilla gardening' action on Mayday 2000
in London and elsewhere around the UK (*Guardian*, 2000d). This event
involved the practice of 'Digging' and remaking space on the symbolic level.
On this occasion Parliament Square was dug up and the road nearby was
'greened' with the turfs taken from the square itself. Incidentally, a banner
bearing one of Winstanley's maxims was strung between lamp-posts on the
square, confirming the link between the action and its historical antecedence.

In similar fashion to the St George's Hill occupation, TLIO repeated a
comparable historical, radical land event later that year in highlighting and
mobilising another piece of alternative history, namely the Norfolk Kett
rebellion of 1549, where enclosures and land rights had also been a central
element of the revolt (see TLIO, 1999). In all, the series of events that have
been instigated by TLIO/Diggers350 can be viewed as examples of a 'postmodern' politics: making use of a range of tactics and mobilisations to promote the overall political aim. The Diggers' actions can be read as a
combination of the historical and contemporary use of, and meanings surrounding, St George's Hill. The use and role of the media was an important
part of the actions in 1995 and 1999, and the media, including a project website, are also playing an important role in disseminating the ongoing memorialisation project (*Surrey Comet*, 2000; see also <www.diggerstrail.co.uk>).

In terms of citizenship and the countryside, the TLIO/Diggers350 campaign can be read in terms of postmodern citizenship. It represents a claim
cast in terms of lifestyle, personal freedom and more sustainable environ-
ments. The claim is underpinned and motivated by historical claims and the
alternative (rural) lifestyles being pioneered by some of its membership;
essentially by practising what is preached. The 'third space' between the
academic and the protest movement came together both in formal terms
through the conference, and through the detailed research and understanding gleaned about St George's Hill. The TLIO campaign and the
contemporary Diggers350 action underline how time can be manipulated
in seeking to rework and influence citizenship construction (status) and can
be influential in terms of citizenship as activity. The mediation and valuation

of history *qua* heritage enables new possibilities for contemporary radical projects to exploit, and, as is explored below, for other individuals to deploy in their narrative translation (see Woods, 1997; Callon, 1986).

The TLIO/Diggers350 campaign has highlighted the linkage between exclusion from the land in a physical and economic sense and also from structures of governance – a problem that the planning system, for example, has struggled to overcome in the past (Fairlie, 1996; Hall *et al.*, 1993). Hence one of the objectives concerns the process of engagement with politics and land management, arguably a central characteristic of participative or active citizenship. By 2000 TLIO had begun lobbying government and other agencies and interest groups on land issues as part of the wider rural/environment/planning policy community. This provides an indication that TLIO (or at least elements associated to it) is moving closer to being a lobby group rather than a direct action movement. This is evidenced by the insider politics being pursued around the permanent siting of the Diggers memorial stone, and the LHI project. In this manner the development, or disintegration, of TLIO may follow a familiar trajectory for environmental pressure groups and the fluid coalitions of active citizens that emerge to contest particular issues (Anderson, 1997; Lowe and Goyder, 1983; Kousis, 1998). This begs the question: what prospect is there for sustained political action on the part of active citizens?

Mobilising the past for the future of the Wye

The Diggers example shows how a political challenge from the past has been used and memorialised in the present as part of a wider contestation of land rights. The next example is taken from the field of environmental planning and politics and surrounds the contested issue of navigation on the River Wye (Penning-Rowsell and Crease, 1988; Penning-Rowsell, 1996). The case study exemplifies how actors can pursue claims in quite different circumstances from the Diggers example. It illustrates how the widened conceptualisation of citizenship employed here can include citizen action in attempts to destabilise dominant definitions. It shows how history and agency combine to challenge nascent rights that, in this case, were close to being authorised by the state.

The focus of the example relates to the actions and tactics used by a particular actor in seeking to challenge the use and regulation of a water resource and environmental asset (see also Parker and Wragg, 1999). This involved challenging rights-claims and promoting extant, albeit 'disused', rights. In this sense there are examples of active and engaged citizenship involved in these processes. Where the Digger narrative involved national-level politics and conflict being played out using the local as a stage, the Wye example differs in that it is concerned with conflict over local practice and rights. Yet it has drawn in national and international issues and actors as part of the argumentation process. Another difference is the approach that the key actors have adopted in the processes of mobilisation and enrolment. I argue that this example is a case where a relatively stable policy community was challenged (Sabatier, 1987; Cigler, 1991) by the promotion of an alternative vision for the River Wye. The environmental groups that had been active in planning and managing the Wye can be seen as a network that had been moving towards gaining authoritative status over river management.

The Wye example connects to the way that particular interests and groups can destabilise networks and power relations. Mandelbaum (1991) sees the approach of the challenging actor as one that, while involving the presentation of alternative histories, is part of the strategy whereby discourses and alternative 'voices' are mobilised as part of 'persuasive storytelling about the future' (Throgmorton, 1992; Macnaghten and Urry, 1998). Where parties are in direct conflict, Grant (1994) sees such an approach as 'competitive storytelling'.

The Wye as multiple resource

The River Wye is viewed as an important resource for both recreationalists and conservationists (see Penning-Rowsell, 1994, 1996; Wye Valley AONB, 1992; River Wye Project, 1992). It is a site of multiple layers, uses and meanings, of which only a few can be mentioned here. The Wye is a source of income for the leisure and tourism industry and for salmon fisheries. It is also an important ecological site. While tensions between economic development and environmental protection are not new, such potentials for conflict and conflicts of interest lie at the heart of land-use planning (see Cullingworth and Nadin, 1997; Grant, 1994; Blowers, 1993; Mandelbaum, 1991). Disputes over appropriate use, development and management of such resources are quite commonplace (Macnaghten and Urry, 1998; Kousis, 1998).

Within such contexts of conflicting alternatives, local authorities and quangos play an important, yet potentially fraught, role as regulators of development, promoters of conservation and also as agents for local economic development (including the development of the leisure and tourism industries). Consequently, they may be placed in difficult situations since the decisions they make relating, directly or indirectly, to the environment may be based on attitudes and information derived from a wide range of economic and cultural sources (see Clark *et al.*, 1994; Mandelbaum, 1991) and subject to differential empowerment. NGOs and local authorities have to balance an increasingly diverse range of interests (local, national, even global) when making political decisions affecting the communities and places that they represent. Grant (1994) argues that as they do so, their script will consist of a mix of discursive acknowledgements of such concerns. Specifically, such tensions are acknowledged in the policy documentation of virtually all statutory bodies whose policies have impacts on the Wye Valley environment and wider locale (see Parker and Wragg, 1999).

The Wye Valley has been a site for an intensification of policy proposals and designations designed to protect, conserve or enhance its amenity and environmental character. The integration of various goals and the need to involve the public in policy-making has engendered a shift towards increased partnership working and 'consensus-building' (Selman and Wragg, 1999; Innes, 1996). There has been an increasing emphasis placed on public participation and empowerment in the policy process (DETR, 1998c, 2000a; Darke, 1999; Countryside Commission, 1998). For example, land-use planning has, importantly, shifted its attentions to some extent to cope with the challenge of sustainability and sustainable development, which in turn has led to a renewed focus on balancing the often conflicting interests of economic development, environmental protection and social inclusion. Such a challenge compounds the difficulties of mediating between such disparate groups in a 'crowded' policy environment and in constituting texts or narratives that successfully reconcile and accommodate such objectives.

In parallel with the increasing recognition of conflict and conflict mediation as being central in resource planning, local authorities and other institutions have been moving towards encouraging local empowerment (Clarke and Stewart, 1998; New Economics Foundation, 1998; Wilcox, 1994). There has also been a more explicit recognition of the connection between people, place and their cultural heritage; the Local Heritage Initiative is part of this. Communities are seen as being an important part of a landscape's resource, and local involvement in shaping policies is becoming a priority (Countryside Commission, 1996a, 1998). Such connectivity between people and place is an explicit recognition of the symbiotic relationship that is part of environmental citizenship. It is questionable, however, whether the type of participation being embraced does represent true empowerment and whether consensus-building does adequately reflect or include all interests and the points that they may raise (Environmental Resolve, 1995).

Within the backdrop of increased public participation, initiatives and exhortations and the restructuring of environmental regulation (Buckingham-Hatfield and Percy, 1998; Goodenough and Seymour, 1999), the heart of the 'dispute' outlined here is the contestation of the appropriate use of the river. There is a shifting dynamic between the river as means of economic intermediary and the river as an environmental resource; it is designated as a Site of Special Scientific Interest (SSSI) and is in an Area of Outstanding Natural Beauty (AONB) (see Wye Valley AONB, 1992). The potential narratives are complex and involve many different groups; one attempt at piecing the history or archaeology of the conflict together is attempted in Parker and Wragg (1999). Here, however partially, the purpose is to illustrate how an actor sought to contest (see, for example, Throgmorton, 1992) the story and regulative controls assembled and proposed by the Environment Agency and other pro-environment actors involved in the Wye Valley (Wye Valley AONB, 1992; Environment Agency, 1997).

The Environment Agency (EA), the body now responsible for water resource protection in England (it replaced the National Rivers Authority (NRA) in 1995), has, laudably, moved towards extended consultation in policy formulation; it actively seeks input from interested parties (Environment Agency, 1998). For example, the Wye Valley Catchment Management Plan draft prepared by the NRA in 1994 received eighty-nine public responses (NRA, 1995) and the Environment Agency's Local Environment Action Plans (LEAPs) involve extensive public consultation (Environment Agency, 1998). Yet this case illustrates how those opposed to even the most inclusive of processes of policy-making can destabilise such plans.

The central issue considered here is that of navigation for boats and other craft on the river and the arguments generated about that in terms of river management. This practice was presented as part of a question and contestation both of legality and of best use of resources in terms of power and utility. The river has a history as a waterway for moving goods and people dating back to before the seventeenth century. Its status as a public navigation was secured by legislation dating from 1662 and subsequent Acts (Stockinger, 1997), although by the mid-nineteenth century its use for trade had declined, and gradually it became less and less navigable. The legal right was still extant, however, and from time to time provided the cause of conflict. For example, the *Ross Gazette* (1994) told how boatmen 'had accused salmon beat owners of blocking ancient rights of way on the River Wye' and were concerned that 'the principle of free navigation on the river, alleged to date from Edward the Confessor, was under threat'.

By the 1980s the river was already under particular informal management regulations, with a series of institutions managing it. As a result, periodic attempts to gain the status of legal navigation authority, or to introduce restrictive by-laws over the river, were being made (MAFF, 1972; Wye Valley AONB, 1992; Parker and Wragg, 1999; Penning-Rowsell and Crease, 1988). The right of navigation was an obstacle to the environmental network, and the threat that people could disrupt the river through their practices loomed over the network. This pertained even though the right was exercised only in a few short stretches of the river and by small craft and canoes. In 1985 an example of inappropriate use took place: a fleet of hovercraft legally navigated the river, much to the chagrin of local figures and the Welsh Water Authority (at the time the 'responsible authority'; see Parker and Wragg, 1999). This event prompted renewed attempts to impose legal regulations on river use (Wye Preservation Society, 1998).

There are several key interests involved in the Wye. Here we centre on the conflict between the boating interest and the environmental interest over the right of navigation. The environmental interest for the Wye was headed by the Environment Agency, which became the main interest opposing the boating interest. The boating interest is represented by the central actor, who used a discourse of economic development to further his case and pointed towards an historical and heritage storyline to underpin the case.

Our 'active citizen' went through processes of counter-mobilisation and counter-enrolment in order to (attempt to) redefine the issue and reconfigure the emphases placed on various priorities (e.g. environment, tourism, fishing, amenity, 'restoration') to accord with his own interest.

In the texts relating to the river's informal regulative regime headed by the EA, three main reasons were given for attempting to restrict use of the river: first, its increasing use for leisure and recreation, with the potential for intra-use conflicts; second, the river's value as a natural habitat, which might be threatened by unmanaged use, including unregulated navigation; and last, the decline of fish stocks (specifically salmon on the Wye), both for commercial fish-farming and for recreational fishing purposes, which should be preserved and enhanced through appropriate management. The approach found favour with many of the interests on the Wye, which were enrolled to imagine benefit from the regime. For example, the fisheries owners saw the prioritisation of the environment as important to ensure fish yields – particularly as the salmon were alleged to prefer shallow water.

Into this relative stable policy environment entered a local entrepreneur in the early 1990s who planned to set up a floating restaurant on the river at Hereford. He bought a 150-ft Dutch Rhine barge (Plate 5.2) in the Netherlands for that purpose and sailed it around the coast and up the Wye in order to realise his plan. His actions, involving navigating the barge up the Wye (with considerable difficulty due to low bridges and shallow water), alerted the interests generally opposed to navigation. He was refused permission to moor the barge in Hereford and was informed by the local authority that he required planning permission for the proposed restaurant and licences for the craft. After this initial setback, the entrepreneurial citizen began carrying out historical research into the history of navigation on the Wye.

He enrolled a maritime solicitor and attempts were made to locate the original shares in the ancient Navigation Company, dating from 1809, which had been the legal navigation authority for the river. It transpired that the company had never been wound up officially. By 1993 the actor had been instrumental in setting up a trust with the central purpose of restoring the Wye to a fully navigable state using the powers claimed under the company. By mid-1994 he and others had attempted to become trustees in the Navigation Company with the intention of 'reviving' it in order to claim the necessary authoritative status. The shares in the company became important texts, resources or intermediaries in the network being assembled by the entrepreneur – this resonating with Latour (1994), who argues that 'networks are thus made up of diverse materials – humans and non-humans – which enables them to endure beyond the present'.

The would-be entrepreneur then claimed the right to manage the river through the legal (or legitimated) navigation authority. He attempted to mobilise further resources in his network in order to claim what Jessop (1990) terms 'autopoiesis' or radical autonomy – essentially the ability to

Plate 5.2 The *Wye Invader* moored on the River Wye near Hereford. (*Photo*: Gavin Parker)

control and regulate the use and management of the river. The entrepreneur attempted through his actions to destabilise an arrangement of political priorities. The deployment of historical resources was both discursive and symbolic, and potentially still held congealed power, or legal authority, that would be problematic for the Environment Agency to ignore. The struggle to control the issue of navigation has since resulted in the need for a series of judicial decisions in which the Environment Agency has sought network 'closure' or authoritative resolution (Bijker *et al.*, 1989; Mandelbaum, 1991).

Mandelbaum (1991: 211) argues that one of the strategic outcomes of the use of 'competing stories, seeking to resolve differences so as to mobilise resources and consent' is that authoritative processes (legal) will resolve disputes and provide an ending. In such a scenario there can be only one historical 'reality' that prevails. Although the courts found against the entrepreneur's claim to have revived the Navigation Company, a public inquiry was held early in 1997 into a Proposed Navigation Order that would enable the EA to introduce by-laws. The inquiry reported later in 1997 but in early 2001 the outcome was still to be decided (Environment Agency, 2001).

Place, destabilisation and reterritorialisation

Such practices may be linked to particular constructions of place. The importance of history as a method of attempting to delineate or relineate the preferred future is considerable. Further, it is argued that the discourse of heritage that prevails with a positive or benign implication assists in this aspect of 'postmodern' politics. The use and deployment of history/time and how differential times are mobilised in the contemporary are part of the mix of fluid and the processual politics of the rural (see Bender, 1993; Hetherington, 1996). It is argued that time is retrieved and made to do service in the contemporary as part of remaking place and progressing claims through and for practice and over space. Law (1994) suggests that a central actor or 'network-builder' governs the combination and redefinition of materials as it seeks to monitor, represent and hold together diverse places and times on its own terms. The government undoubtedly realises how momentous the decision will be for the Wye and dependent activities.

The Wye narrative shows how spaces cursorily regarded as 'prescriptive' leave room for negotiation or challenge (Murdoch, 1998). The challenging actor was seeking to reterritorialise the space of the Wye using diverse resources including the boat and the historical legislative texts. It is argued that not only is the rural fluid in the sense of contingency, but also the notion of liquidity and the inference of liquidity, being flat or even, should be treated carefully. Such a synthesis can lead to an understanding of rural space as being uneven, multiple and fluid, where culture and practice vary from place to place, group to group and time to time. When actors attempt to exert power they draw upon resources within those variables, looking also to fracture conceptualisations of place and practice and to 'make room' for their own synthetic, persuasive storytelling. In this way examples of truly political, active citizenship can be discovered.

One stratagem explored here is the use of history to challenge extant conditions. The representation of heritage and space and the cultural imagination can be used to express the political. In this way, time and current practice (standing conditions) are challenged by presenting alternative practices or 'stories'. The examples discussed above exhibit similar general characteristics in the sense of protest: the claiming of rights and the enactment of past practices in particular 'action spaces' or places. Such examples represent part of postmodern politics that is multigrained, often fleeting and relating to a single issue and yet diverse in terms of objectives and tactics employed, as well as being often reflexively part of a wider process of resistance. Tactics used include performances; deploying ironicism, humour and irreverence; and seeking to appropriate opponents' discourses in order to ridicule or subvert them (see, for example, Chesters, 2000).

In the context of this volume, such actions and reactions can be read as plays over rights-claims and counter-claims; therefore as aspects that might restructure trajectories of cultural fields. The claims and stories told by claims-makers are portrayed as clear-cut and incontestable. In reality they may be shown to be messy, partial and 'leaky'. Actors both attempt to convince others of the legitimacy or authenticity of their claim and attempt to undermine other claims. Resistance to dominant narratives may lead to practices played out in spaces and places that may of themselves be histor-ically significant or resonant and/or tied through space and time. As a corollary, citizenship development within that locality, and indirectly nation-ally, may be influenced.

In the Wye Valley example the action was tied to the River Wye specifi-cally, and the claims being made were in that sense bounded and 'local'. In the Diggers case the site was selected for its 'heritage value' and as an action space to be exploited to pursue a national campaign. It is also the case that the use of history/heritage may be discursive in the sense of textual deploy-ment. It may be that the action rather than the space/place dictates how and where the action takes place and therefore forms both the content of the action and the space in which the action takes place. However, the asso-ciation of claims to places also appears critical to the pursuance of the claim. For our examples, both alternatives are represented.

Conclusion

The actions outlined illustrate attempts to rework modernity by rendering such histories in non-linear fashion. Therefore the notion of multiple time–spaces is drawn into this wider discussion of citizenship and land (see Laclau, 1990). History or 'old time' (aspects, fragments or selected read-ings) are used and deployed as part of political projects with the effect of destabilising relations, cultural arbitraries and, sometimes, physical environ-ments. In one sense this 'raiding' of history is done with the intent to remake space and reconstitute society along the lines preferred by the mobilising actors. Such deployments may be discursive in the sense of textual deploy-ment, or may also lead to practices played out in spaces and places, practices that may of themselves be historically significant or resonant and/or tied through space and time.

While it can be said that the Wye example is of the local, it draws on and deploys regional, national and even global concerns and resources. It provides an example where a group uses a local and ancient action as a beginning point, or lever, to progress a wider campaign to reorganise and redistribute national and even international (land) rights and aspects of policy/decision-making. The Diggers episode represents a good example of drawing upon history, particularly 'heritage', for such a purpose. The latter part of the story also shows how selected past events and practices can be adopted and even celebrated in the contemporary as part of a postmodern *bricolage* of cultural heritage. It is arguable whether the portrayal and signi-fication of the Diggers sites will succeed on its own in effecting political change, but it is one of a range of tactics that can be used by claims-makers, as argued in Chapter 3.

With the Wye example, the key actor was operating from self-interest in the first instance, bringing or enrolling texts and intermediaries bearing particular discursive stances or resources that could be deployed in his cause.

His story was one of the river as both an environmental resource and an economic one; in recounting its history as a waterway and by attempting to reanimate old dispositions of power he claimed to be able to control the management of the river. The story told by the actor was bold and the example is perhaps an extreme example of dissonance and engagement on the part of a citizen.

Law (1999) talks of relational materialism; basically, how and why things get together or are assembled by network-builders (Selman and Wragg, 1999; see also Woods, 1997). Evaluation of how such building is achieved might be carried out through a closer investigation of the contents and processes in further, future research examples. The examples given show how citizens, in pursuing alternative claims (particularly ones that conflict with dominant views of space and practice), manage, or attempt to construct, their challenges and narratives. In these processes, related in albeit partial accounts above, there is a question about the ability of groups such as TLIO to extensify the engagement of people in political discussion and action in relation to land use. This question mark remains despite the example of the Diggers memorialisation, which appears to have enabled such a process in at least one locality through the use of a legitimating discourse of heritage. It is clear that much more detailed, and possibly longitudinal, ethnographic work is required to unpick the detail of how and why citizens engage, or not as the case may be.

Criticisms have dogged environmental groups that they have been elitist. TLIO as an organisation made attempts at its inception to be internally democratic and 'non-organised' (TLIO, 1995). The paradox of this is that many issues tackled by environmental groups do require certain levels of information and knowledge that are shared between networks. Wider groups of citizens not only lack knowledge or information but also rely on specialist groups to pursue particular issues. Groups using direct action, in particular, also suffer, as actions of this type are usually planned by a small group, and often secretively in order to resist infiltration by forces of 'law and order'. The use of the Internet may alter this, but even with experience to date, key information is released only shortly before an action takes place, thus limiting dissemination (Pickerill, 2000; Mobbs, 2000).

In this respect, land and property rights issues can appear to be daunting and complex issues to engage with as well as being rather obscure or esoteric to most people. Land and rights have not had a high profile, or been at the forefront of (English) political culture. Accordingly, TLIO might be more usefully seen as a catalyst, and might view its actions as mediating (and mediated) actions carried out in order to raise awareness and interest among a wider public. In this sense the occupations of St George's Hill may be viewed in this way: active citizenship with an aim to change structures, but also with an educational purpose. It should be mentioned that land reform and property rights issues reached the formal political agenda in Scotland with a bill to reform Scotland's feudal land system being introduced in 1999, as one of the first pieces of law to be put before the new Scottish Parliament (Scottish Executive, 1999). Even though this has come under criticism from some quarters (Fairlie, 2000; Gibb, 2000), people appear more informed and active about land issues and are engaged in criticising the proposed Scots legislation. Given the series of recent rural/global crises, particularly those relating to agriculture, structural economic issues such as land use and reform may become more open for discussion and alternative strategies propounded.

On the level of contemporary rural change, such events and examples relate to the way that regulation and cultural norms extant over rural space are uneven and contested. It is possible that such tensions are rendered more acute by changing land use, the need to diversify land use in the country-side and counter-urbanisation (see Halfacree, 1996; Boyle and Halfacree, 1998), which add to the potential for instability in social relations in rural localities and may give rise to conflict. In the case of the Wye entrepreneur his attempts to bring in others for legitimation and validation was, and will be, secondary to his original motives based on self-interest. It appears in both examples that strong, motivated and informed individuals remain the key to challenges to dominant power.

Chapters 6 and 7 pick up on other expressions of citizenship: as primarily state-conceived active citizenship and consumer-citizenship respectively. In particular, examples are used that have related to land and its use and are set within a rural context, but others illustrating a wider range of 'citizen' actions are mentioned. In Chapter 6 the way that the state and dominant or 'embedded' power reacts to and uses active citizenship rhetoric is examined. This perhaps illustrates another extreme from the examples used here: where the state has attempted to activate citizens, albeit boundedly, in particular projects to further government policy in some form. In Chapter 7 the proliferation of citizen labels and types, and their mode of expression, scales of engagement and the issue of political passivity, are assessed in relation and juxtaposition to the dynamics of rural change. A discussion of the need to amend political structures and to harness the energy and commitment shown by activists and protesters, and yet unexpressed formally by the vast majority of people, is included.

6 Political expediency, localness and 'active' citizenship

Introduction

This chapter picks out examples of the micropolitics of the rural, but in contradistinction to other chapters focuses on examples of citizen action and policy that has been explicitly directed towards citizen action by government. This is often expressed through attempts on the part of the state to encourage particular forms of active, 'good' citizenship. This chapter therefore details a policy vehicle, the Parish Paths Partnership scheme (P3), aimed explicitly at engendering rather anodyne 'active' citizenship, and relates the outcomes associated to the scheme. This vantage of citizen action contrasts to Chapter 5, where the examples were largely mobilised by particular and marginalised actors in directly challenging authority. Drawing on and adding further conceptual detail of the theoretical position set out earlier, I argue that citizenship's aspects of status, identity and activity are interlinked and processual. Connections are made here between the theoretical concepts introduced and extant rights distributions and policy instruments in the countryside (see Parker, 1997; Munton, 1995; Whatmore *et al.*, 1990) and consequently how such processes and flows of power are linked to 'a politically inspired restructuring of human rights' (Smith, 1989: 148). It is asserted here that a key aspect of governance and rural governance is the trade-off of rights and responsibilities that takes place between individuals, groups and institutions.

Aspects of citizenship activity linked to identity and status definitions can be discerned or read in numerous contexts and practices, particularly when looking to reinforce specific aspects of self-image or to gain prestige. Citizenship as discourse can be adopted easily by governments and drawn into widely disparate projects. UK governments over the past twenty years (and before) have used this approach to legitimate policies and to galvanise the population actively to do something for, *inter alia*, the 'community', the 'nation'. Public participation in aspects of rural policy can thus be viewed or represented as 'active' citizenship even where such 'citizenship' is manipulated and controlled by the state. Governments have recently expressly invoked citizenship as a means of achieving objectives and have used participation as a condition for funding numerous projects. Through the analysis of one such example here, the issues raised by attempts to engage (and be defined as 'engagement') in policy formulation or implementation are set out. Links are drawn to issues of identity-construction and status-building, and issues concerning contemporary measures encouraging empowerment. The discussion leads into the consideration, in Chapter 7, of alternative practices of participation and 'citizenship' and 'consumer-citizenship' that have been features of rural/environmental politics during the 1990s (Parker, 1999c).

Modern state, postmodern citizenships?

Some commentators on the citizenship debate feel there has been such a shift in and fragmentation of (global) society that we can think in terms of 'postmodern' citizenship (see Urry, 2000). In this context, citizenship as status and as label might be understood as an 'authoritative resource' (Giddens, 1984) to be drawn upon and capable of becoming yet another capital commodity. It can be drawn upon by individuals, partly through their participation in certain activities and subsequent use and deployment of such activity/participation, during the course of their life. It may also be adopted and drawn upon by group interests (see, for example, Bourdieu, 1990; Clawson and Knetsch, 1965). Such a use of citizenship, however, implies selectivity and the need for political/cultural consciousness, as well as the ability to express opinion through behaviours in different spheres – as mentioned, a degree of cultural competence or perhaps social capital (see Fine, 2001).

One of the other key features of a fragmented 'postmodern' citizenship is the emergence of more overt competitive rights-claims and citizenship claims by individuals and groups. As part of the process of making a claim over (the distribution of) rights and responsibilities, new forms of citizenship are of themselves created – nascent citizenships fostered by a mushrooming (and withering) of sociations, communities of interest and different practices. Citizenship becomes a processual and practised role signified by different spatial and discursive practices. Citizens act inconsistently and people express 'their citizenship' differentially. Following from the broader understandings of citizenship described earlier, it is asserted that through the study and analysis of micro-political action it can be discerned how seemingly inconsequential actions are building blocks of citizenship construction and maintenance – of the practice of citizenship. Obversely, the state acts unevenly and inconsistently in terms of regulating people and spaces (see, for example, Keith and Pile, 1993), and it is reasonable to suggest (though not necessarily in correspondence to the above) that different spaces are differentially subjected to state regulation (see Rabinow, 1991: 243). Citizenships rely less and less on the nation-state and more on local networks and extra-national relations of power. Through the process or performance of action and interaction, rights and responsibilities 'become' – allowing for multiple domains or 'communities within community' to exist and vie for reification, or perhaps more specifically valorisation (see Pred, 1984; Kristeva, 1984; Raban, 1988; Ravenscroft, 1999; Etzioni, 2000).

As part of the restructuring of social and political relations between classes, groups, institutions and business (see Giddens, 1991), there have been subsequent and predictable reflections of localism and individualism in countryside policy (see Goodwin, 1999; Parker, 1999b). As discussed in Chapter 3, this was perhaps particularly the case during the Thatcher and Major years, with, however, seemingly little change under Blair. Such reflections are widespread, from the national political policies of relaxing planning controls to the use of citizenship rhetoric and the attempts of politicians to shift state responsibilities outwards – a form of hollowing out of the state rather than a rolling-back (see Jessop, 1990; Goodwin, 1999). Whatever the tool or vehicle being used, it relates to citizenship and participation (or 'practice'; see, for example, Crouch, 1999).

Rights are conceptualised here to be somewhat differently interpreted or enforced in local situations to those recognised by the state. Rights can develop through their formulation at the local scale in a bottom-up way and may be formalised and given a universal sanction by the state, becoming an authoritative resource. In similar fashion, rights given legitimacy by the state can be redefined at the local level in an uneven dialecticism, or a form of 'dialectic of practice' (Bourdieu, 1990). The nation-state is less able to formulate and regulate the ingredients of that form. Attempts to gain dominance over local social relations are made by powerful individuals or groups or may be eroded by practice over time (the creation of rights of way through immemorial use being a classic example; see Bonyhady, 1987). Rural areas are often problematic for the state in terms of surveillance and the enforcement of legal rights and responsibilities, and the way in which legally defined rights and responsibilities are observed at the local level is likely to vary according to local cultures and be susceptible to local interpretation. As a result, discursive battles are waged, or coalitions made, between different scales of interest (for example, national media versus a local group), attempting to shift definitions of legitimate activity in any given place, or over any particular issue.

Rights are examined in parallel with the relationship between agency/structure interaction as examined by Bourdieu and others (see Ritzer, 1996; Jenkins, 1992). For example, Giddens's (1979, 1984) theory of structuration is broadly comparable with Bourdieu's dialectic of practice in relation to agency–structure interaction. The conceptualisation of structuration given by Giddens (1979: 66) involves 'conditions governing the continuity of structures, and therefore the reproduction of systems'. These are formulated by a variety of actors on different levels of operation or according to the contingent or episodic power relationships present (see also Clegg, 1989). There is a dialectical relationship between contested authority and identity in the countryside. Within this interactionism, rights and socially sanctioned activities *qua* rights (Batie, 1984) are conceptualised as being a product of such interactions. Similarly, nascent 'rights' or implicit responsibilities are regarded as being manifestations of localised power relations and results of activity that groups or individuals engage in. These may or may not be reflected at the level of the state or by powerful groups who may influence rights distributions through policy and legislation (see Van Gunsteren, 1994). Commentators such as Crouch (1999) see everyday practice as being embodied relations of the rural. They are constitutive as well as products of relations of power – presenting opportunities for spontaneity and resistance as well as being determined by internalised authority.

Local empowerment and citizenships

In addition to theories and possibilities of citizenship, arguments over the centralisation of power and debates over local empowerment and allied devolvement of power and responsibility have been features of UK politics over the past ten years at least (see Burns *et al.*, 1994; DETR, 1998; Clarke and Stewart, 1998). Given the 'New Labour' commitment to decentralisation and devolvement of power on the part of the Blair administration, such processes are likely to be pursued further. As a logical outcome, for example, parishes are likely to receive opportunities to extend their community governance role, and 'active citizenship' at the local level is likely to continue to be encouraged (Institute for Citizenship Studies, 2000; DETR, 1998a). In England, however, this is likely to remain at the level of handing down responsibilities and perhaps allowing some limited powers to the parish councils or encouraging local authorities to 'allow' people to input into their policy-making processes, thus little altering the dispositions of power (see Clarke and Stewart, 1998).

As has been discussed in earlier sections, recognition and discursive exhortation concerning 'active' citizenship and community 'empowerment' have multiplied during the 1990s. This is due in part to the increasing use of the term as a representational tool and a rhetorical or discursive device and also as part of a move to look at different ways of engaging people in decision-making processes and other public activity. It is anticipated that efforts to 'get closer to people' – to empower at the local level or to create strengthened 'community governance' – will continue to be encouraged by the state and will be demanded by a variety of sections of the population at large (Institute for Citizenship Studies, 2000; Davies, 1999; DETR, 1998a; Stewart, 1995; Gaster, 1996; Murdoch, 1997a; Giddens, 1991). Conversely, other groups and individuals may resist such moves. Such systemic change may give rise to its own tensions as agents attempt to rework and create their own 'spaces of control' (Giddens, 1982). However, 'citizenship' is likely to remain a power word that will be deployed by the state and by individuals to legitimate policy and practice.

Preconceptions of the 'rural' and dominant cultural imagery have historically played a part in policy formulation for rural areas. The 1995 rural White Paper for England (DoE/MAFF, 1995; and as highlighted in Chapter 3) says much about the previous Conservative government's attitude and intentions towards parish empowerment. The paper can be seen as an incorporation of certain internal (and enduring) assumptions concerning rurality and 'rural people'. For example, it supposes that rural communities have 'traditional strengths' which are 'communally active enterprise and voluntary activity' (*ibid.*: 13). The government played on the suggestion that rural areas are examples where it could be 'encouraging active communities which take the initiative to solve their problems themselves' (*ibid.*: 10). Again this type of rhetoric is echoed in the 1997 Local Government Rating Act, which concerns parish reviews: 'it is desirable that a parish should reflect a small, distinctive and recognisable community of interest, with its own sense of identity. The feeling of local community and the wishes of local inhabitants are the primary considerations' (DETR, 1997: 2).

Such sentiments are unlikely to be reshaped dramatically in policy terms (see DETR, 1998c, 1999; Clarke and Stewart, 1998); the image of the rural is both convenient and simple for governments to deploy. There is a growing literature that points, however, to difference or 'multiple countrysides', and serious socio-economic problems and barriers that lie beyond this still dominant rose-tinted view of the rural (see, for example, Cloke and Little, 1997; Milbourne, 1996). While an extensive review of such literature is not possible, it is important to problematise the underpinning political view taken of the rural and hence of the policy under scrutiny. Here, albeit through a political economy lens, citizenship is used as the discursive means through which to explore and link discourse from government, policy-makers, and from people who live, work or take their recreation in the countryside.

There are a series of questions that can be constructed to prime attempts to interrogate citizenship. How does policy aimed at engendering and strengthening citizenship and associated notions of community impact on contingent local relations of power? (see Flyvbjerg, 1998). How does policy that helps shape citizenship *and* understanding, or reflexive understanding of citizenship, provide for conflict as networks attempt to integrate or repel alterations to stabilised relations? (see Parker and Wragg, 1999). How does policy alter practice, in both the short term and the longer term?

Citizenship and contesting countryside access

Attempts at blanket prohibition of transgression against rights distributions reinforce (hegemonic) standing conditions that favour powerful groups and private property owners, thus further distancing the rights-claims of other societal groups. This reinforcement is partly achieved by manipulating the social context in which political contestation occurs (Whitt, 1979). Such manipulations have meant that many dispositions of power have remained in place despite occasional challenge. The P3 scheme provides an example; it is explored below. Rights become indistinct as they are appropriated and obscured, or championed and derided by particular groups or individuals. A clear case in point in this respect is the issue, definition and practice of 'trespass'. Trespass and transgression is a theme that is carried through into this chapter, picking up from aspects that were explored in Chapters 4 and 5.

'Official' or legal rights and responsibilities are clearly not adhered to evenly in space and time, and cultural practice and relations of power reinvent and appropriate spaces. In some cases other groups rely on legal structures to support their view of citizenship. In this sense, rights and responsibilities in the 'hard' or juridical frame can be seen as attempts to punctualise or routinise culture and practice. This situation provides classic 'terrain' to examine citizenship in terms of activity and identity construction. Such relationships and inductive hypotheses are best reviewed where differing conceptualisations of appropriate activity and rights/responsibilities are expressed over a particular space.

Countryside access and rights of way, in particular, can provide interesting cases for study (see Crouch and Matless, 1996; Ravenscroft, 1999), not least because there are defined legal rights and responsibilities set out as regards who should do what and where they should be doing it (see Countryside Agency, 2001). This is often in stark contrast to what people actually do (see Uzzell *et al.*, 2000). Both legal and illegal practices and policies give rise to many interesting conflicts and political manoeuvrings at the national and local scales. The examples used here relate to the problems of enforcing the law and legal constructs of space and practice and how such differentials allow for different interpretations and constructions of rights and responsibilities to emerge. The Parish Paths Partnership scheme (P3) is used to exemplify the way that rights and responsibilities are contested and traded 'on the ground' and envisaged and enforced at different scales. Discussion of the scheme includes consideration of how local relations shape rights and their interpretation; where alternative conceptions of rights/responsibilities are presented there is often observable conflict (and, perhaps even more often, less visible friction).

It is illustrated here that policies cannot be fully assessed without recourse to the cultural environment in which they are deployed and therefore are dependent on the forces that are at work to allow or distort policy intentions, through human agency, local culture and appropriation. Outcomes are dependent on the mediation and distribution of rights and responsibilities, and the way in which rights and responsibilities are enforced so that citizens are enabled to participate effectively. This latter point is central in terms of discussing particular local resistance to cultural change. Both points are important in the general analysis of the way that environments are read and structures altered and reinforced; examples are used to assess citizenship theory in the countryside. The case studies that follow are drawn from research into the Parish Paths Partnership scheme (Parker, 1997, 1999b; see also Goodwin, 1999; Countryside Commission, 1994). Both relate in some way to the P3 scheme requiring participation in countryside policy and planning. They contrast to the examples used in Chapter 5 in that they relate to conflicts over the management of countryside recreation. They are examined to illustrate how people become animated over rights and the manner in which rights are handled, enforced or observed locally; how local standing conditions determine actions; and how citizenship is contested through practice/activity. The examples detailed aim to illustrate how policy and action relate to and create citizenship (citizenship as understood as being both processual and providing/being sustained by status, identity and activity). Rights of way (ROW) in England are often literally as well as abstractly subject to these outcomes. In this context a right is what is understood as being a right; as what is understood in terms of the practical consciousness of the agent(see Healey, 1997; Giddens, 1984). It may be viewed as a rule that is interpreted within its context (Clegg, 1994) or its domain (see Kristeva, 1984; Donzelot, 1980). In the case of the P3 scheme it was the government and the Countryside Commission that seized upon the idea of local community participation in aspects of land management (in common with other similar schemes; see Goodwin, 1999) to carry forward the rhetoric of citizenship, active citizenship and good citizenship.

Citizenship and activity: (re)mapping and weaving

P3 has been selected to do service in illustrating state attempts to deploy citizenship rhetoric and voluntary action. It also carries a reading of sociopolitical events and processes in the English countryside and British politics more widely (see, for example, Goodwin, 1999). There have been related countryside projects that share some similar characteristics such as Rural Action (Martin, 1995) and Parish Maps (Crouch and Matless, 1996) inasmuch as they involve aspects of voluntarism and local participation. The P3 scheme presents itself as an opportunity to assess how empowerment/citizenship rhetoric contained within government policy is translated and received at the local level. It is through the examination of several texts aimed at revealing the P3 scheme that various aspects of citizenship as posited are explored in the rural context. Work carried out for the Countryside Commission (CC) by private consultants to assess the scheme is used (PACEC, 1995) and personal research is deployed (Parker, 1997, 1999b), and other official texts that have promoted or highlighted the scheme are deconstructed.

These sources provide horizontal or layered inter-textual analysis. Additionally, and horizontally, two different takes on the scheme are set out in narrative case studies, in similar fashion to those of Chapter 5. The first looks at how other 'active citizens' involved in countryside management reacted and contested the application of the scheme within the Cotswolds (AONB) and the second charts how one local P3 group encountered difficulties with local landowners in attempting to implement the scheme in several parishes in north Gloucestershire. In applying and relating the theoretical elements discussed earlier, I provide examples of how citizenship has been deployed discursively from above, how it is received, understood and used on the ground, and how contestation arises in terms of rights and responsibilities.

Context: P3 and its masters

In 1987 the then Countryside Commission made a pledge to open up the whole of the rights of way network in England and Wales by the year 2000. It was this aim that underlay the inception of the P3 scheme in 1992. The target of opening up the rights of way system implied by the National Parks and Access to the Countryside Act 1949 (see Blunden and Curry, 1990; Parker and Ravenscroft, 1999) had historically been a problem for most local authorities (Countryside Commission, 1987, 1993a; Parker, 1997; Curry, 1994). Two of the main factors that have obstructed the policy of opening the whole rights of way network have been persistent resource problems and the obdurate non-cooperation of many landowners in fulfilling legal responsibilities towards rights of way across their land (see Curry, 1996; Shoard, 1996). This latter point is, of itself, important in this exploration of different types, levels and conceptualisations of citizenship in the countryside. Given such problems, the P3 scheme was designed to get people involved in working on their local rights of way as a means of overcoming the difficulties experienced by local authorities trying to open them.

The scheme was funded by the CC, and each highway authority that joined administered P3 within its jurisdiction. The rubric of the scheme allowed individuals to join P3 on behalf of a parish. Therefore the parish council or similar did not necessarily have to be involved or in agreement about scheme entrance. This was a prima facie indication of potential frictions within participant parishes. The P3 scheme allows parishes to carry out work on the rights of way network within their parish. Work was intended to be done to agreed specifications and always to be exactly on the line of the right of way (rather than skirting or ignoring problematic areas). In this way the definitive map setting out rights of way was to be observed (Crouch and Ravenscroft, 1995; Bonyhady, 1987). Importantly, P3 operated with four stated objectives (Countryside Commission, 1994: 1):

- to establish a more efficient, effective and economic means of keeping the rights of way [ROW] network open and in use;
- to allow highway authorities [HAs] to concentrate on matters more appropriate to their role, responsibilities and expertise;
- to unlock hitherto untapped resources at the local level; and

- to enable local people to make greater use of the rights of way in their areas.

The objectives of the scheme can be read as framing statements for good citizenship and justificatory parameters for the devolution of state obligations. More specifically, critical issues are raised by the scheme's objectives: just what resources are untapped? To what extent is it legitimate to offload legal responsibilities from the state or private landowners onto the public? What constitutes more 'appropriate matters' for local authorities when rights of way enforcement and management fall largely within their ambit already? There are numerous surface claims made about community participatory mechanisms including capacity-building and social capital-raising (Falk and Kilpatrick, 2000), but the increased use of public volunteers is terrain that should be contested. Much of the notion of active citizenship rests on an assumption that such 'activity' (or its underlying prompts by authority) is relatively unproblematic in moral or cultural terms (see Selman, 1996; Marvin and Guy, 1997).

The Countryside Commission monitored the scheme between 1992 and 1995 through a private research consultancy (PACEC, 1995). The Parish Paths Partnership final evaluation report used data gathered from monitoring surveys of the scheme undertaken in that period. The evaluation gave clear messages that the scheme had been largely successful in its objectives. These conclusions were arrived at through a normative survey of quantifiable outcomes – for example, 'how many miles of rights of way were opened' or 'how many person-hours were expended'. The evaluation report did not cover who was involved and the effects that the scheme had on volunteers or other people in the parishes involved. It was clear that the report did not tell the whole story about the P3 scheme's effects, outcomes and failings. However, the report remarked briefly and uncritically on the impact of P3 on volunteers, saying that 'Possibly the greatest success has been seen with regard to tapping local resources by stimulating additional voluntary activity within the parishes' (PACEC, 1995: ix). This shows how the framing exercise placed onto the scheme by the CC ensured that the consultants' remarks centred only on the 'tapping' of resources – a seemingly positive wording.

This gives rise to another set of issues in terms of how policy and action are assessed or framed. This (partial) representation of events is of course the stuff of 'spin' politics. Callon (1980: 198) argues that generally 'an initial frontier is traced between what is analysed and what is not, between what is considered relevant and what is suppressed, kept silent', and further, that 'protagonists are involved in a never ending struggle to impose their own definitions and to make sure that their view of how reality should be divided up prevails' (*ibid.*: 207). Goodwin (1999) alludes to this, noting that different policies 'allow' people to participate within certain limits and that such participation determines outcomes and experiences. This point may be true to an extent and it is certainly the case that the individuals involved in activity such as the type being analysed here may be boxed into a particular event-horizon.

The P3 scheme was hailed as a success by the then Department of the Environment (DoE) (now the Department of the Environment, Food and Rural Affairs, DEFRA) and the Countryside Commission (Countryside Commission, 1998). It was explicitly marked out as a 'good example' of local community development within the 1995 Rural White Paper (DoE/MAFF, 1995). P3 was highlighted thus: 'we commend the effectiveness of the Parish Paths Partnership initiative, but we also wish to identify further ways of encouraging direct management at the local level' (*ibid.*: 10). This statement deserves further attention. For the government of the time, the scheme represented a means of shrinking the state in terms of using less public funding (or hollowing it out) while using the legitimating discourse of empowerment and local devolvement. In this way such schemes were being used to justify the devolution of 'responsibility' and 'management' (read work) down to the local/community level, but also relied on the presumed plentiful and willing voluntary brigades that were to be found in rural areas.

The policy environment and the political structures that generated P3 lent legitimacy to the extended use of voluntary workers. In interview, one of the architects of the P3 scheme stated how the climate of policy was right for a scheme such as P3:

> 'Their idea [the Department of the Environment] was that they wanted to empower local people. P3 matched their aspirations . . . we were in the midst of developing policies for community action and rural action in the late 1980s so we did have a suite of policies on community action. The ethos was that way. The whole trend was, and still is, either to empower or devolve responsibility.'
>
> (Countryside Commission officer, 1995)

In this case it seemed that the government and the CC were interested in displaying how 'successful' the scheme was, how the CC had implemented it competently, and how the scheme could be used as evidence of the preferability of such approaches in delivering 'public' services. All three of these could be effectively conveyed using the rhetorical device of citizenship – and, specifically, 'active citizenship'. The explicit objectives that the Countryside Commission laid out for P3 were used to analyse the scheme's performance. However, as with many policy initiatives, it is more difficult to set out the less tangible or non-quantifiable benefits or disbenefits that can be engendered. Indeed, many reviews of policy fail to address some of these non-identified or unproblematised outcomes, or the externalities that can result from the implementation of such policy. Some such externalities can hold significant meaning for attributions regarding the success/failure or the legitimacy of a particular policy. It is argued that by using the theoretical frame set out it can be seen that the assessment of the P3 scheme was partial; and further, that such struggles over problematisation are happening on the ground within other localities and probably for other schemes. The approach taken undoubtedly affected the way in which P3 was prepared and then portrayed. A more detailed study of how the scheme was received and operationalised in one area is set out by Parker (1997, 1999b), and two examples that view P3 in more detail are discussed below. First we examine the way that the scheme was received in the Cotswolds, and second, we see how a particular parish in north Gloucestershire experienced the scheme and the 'citizenship' it engendered.

The P3 scheme in Gloucestershire

Two examples from research in the English county of Gloucestershire are set out to illustrate network conflict in the context of the P3 scheme. The examples are then discussed against the theoretical work outlined. Gloucestershire County Council joined the P3 scheme in its first year of operation (1992) and by 1995 boasted 103 participant parishes, or 12 per cent of the total from 919 participating parishes (across twenty-eight participating local authorities in England). Hence Gloucestershire provided a useful sample to illustrate the characteristics of the P3 scheme as it was practised and resisted.

The Gloucestershire P3 research unearthed some surprising discoveries given the claims made by the DoE and the CC. It was found, for example, that over 70 per cent of the work done under the P3 scheme had been carried out by paid contractors in Gloucestershire – thus introducing another aspect of 'non-empowerment'. This rather subversive strategy of implementation was often achieved by enrolling a P3 co-ordinator within a parish to act as a figurehead. This enabled the county council legitimately to send in a paid work team to carry out specified works. The county council officer in charge of the scheme explained the high proportion of work done by contractors thus:

> '[G]etting work done to a good standard is slightly more important than the slightly woolly idea to get voluntary labour ... we have one contractor ... he followed on from supervising the Enterprise Team, he's done probably 80 per cent of the contract work. He's probably subcontracted some of that, though.'
>
> (Gloucestershire County Council officer)

It is worthwhile noting how the 1995 report was again unable to provide this type of data. This brings into question the need for a scheme at all and also raises familiar questions about the scale and 'buy-in' of the population at large to this type of policy and to the activity *qua* citizenship desired by the state. There are mitigating factors associated with the large-scale use of contractors in Gloucestershire, including the practical difficulties that can be involved with community empowerment initiatives. For example, some parishes simply could not raise enough volunteers for the necessary work, and some of the work was too physically demanding, skilled or complex for the available volunteers. These provide partial explanation for the heavy use of contractors, but these explanations also lend weight to questions about both the legitimacy and the practicality of relying on volunteers to carry out such tasks. Also, the ability of such schemes to build social capacity/capital and the form or content of that capacity should be problematised (Isin and Wood, 1999; Hutton, 1997). The problems with the scheme also add to critiques of top-down citizenship models and the resistance of localities to changes in material and cultural practices. The P3 scheme might have been better received and more truly presented an empowerment opportunity if each parish or area had been able to draw up its own plan for rights of way – but perhaps that would have been too radical and been open to hijack from a group or individual with a vested interest in keeping rights of way closed. It would almost certainly have meant a highly differentiated rights of way policy across England. However, the county council effectively subverted the intentions of the scheme in order to 'get things done', pointedly ignoring the empowerment rhetoric.

Culture, field and resistance

In Chapter 5 the Wye case was used to illustrate how different actors resist attempts at policy-framing and, more generally, how policy itself can represent actual and attempted alterations to rights and responsibility distributions on different scales. Such resistance and alterations may also be seen as an outcome or objective of 'network destabilisation' (see Parker and Wragg, 1999; Callon, 1986), or attempts to rework power relations and material practice over resources. Inevitably, new policy will impact on social relations in some way (see Van Gunsteren, 1994) and has the potential to impact on local culture. In this case, pre-existing relations among local people involved in countryside management and rights of way are viewed through a case study. The Cotswolds Wardens are invoked as examples where attempts to broker change and new forms of participation through P3 were met with resistance by other 'active' citizens. It is argued that the activities and the conflict between the institutions and individuals are themselves constitutive of citizenship.

In Gloucestershire a number of parishes had openly resisted entry into the P3 scheme, and other parishes had faced problems from within their own community as a result of joining the scheme (Parker, 1999b). In particular, problems were experienced within the Cotswolds Area of Outstanding Natural Beauty (AONB). The AONB has its own countryside service and a volunteer section as part of this service, known as the Cotswolds Wardens. The wardens had assisted the county council with countryside management since 1968. Prior to the inception of P3, the service had established its own methods and rationale of countryside management within the AONB using the wardens. These practices included deploying the wardens in the upkeep of rights of way in the Cotswolds.

The management of the land within the AONB had proceeded over time, developing its own cultural arbitrary, a culture that bred particular rules and practices: Clegg's (1989) 'standing conditions'. These conditions related to what paths were kept up, which others were ignored, and what tasks were undertaken and which were not – in short, how the definition of rights and responsibilities on the ground was moulded in practice and in material terms to suit the power relationships present within the social field; how the locality was imagineered or 'mapped' by the wardens and the service; and therefore how the Cotswolds were spatialised (see Lefebvre, 1991). Such a map of order and of relations might be open to constant renegotiation and tactical manoeuvre, but similarly, resistance is encountered in processes of challenging dominant or extant topographies of power (see Murdoch and Pratt, 1997).

When P3 was introduced in Gloucestershire, the wardens were upset that this 'status quo' in relation to countryside management in the AONB might be liable to disruption. In interview, the AONB countryside service manager explained the initial reaction to the P3 scheme in the AONB:

> 'There are politics involved; the Cotswold Wardens felt they had their feet trodden on by the P3 scheme in the Cotswolds because they had traditionally patrolled the paths . . . the local volunteer wardens don't want the scheme . . . there has been deliberate avoidance because of the Cotswolds Wardens . . . we [the Cotswolds Wardens] could pull out literally overnight, which is what we did. It took the P3 scheme a long time to get going, it left a hole – a vacuum – we then had virtual warfare between some of the P3 co-ordinators, the county council and ourselves at the time. In many ways we thought that it would be sensible not to have P3 in the AONB.'
>
> (Cotswolds Countryside Service officer)

The trenchant tone used by the officer reflected the expression of opposition to P3. There was a clear issue about how social relations in certain parishes might be affected by the scheme and how it was felt that the withdrawal of co-operation on the part of the wardens would in some way create a 'vacuum' or perhaps a 'no man's land' drawn between competing topophilias.

The result in the period 1992–95 was that 26 parishes from within the AONB joined the scheme, from a pool of 120 parishes (see Parker, 1999b). The more rigid approach to the rights of way network that the scheme purported to require seemed to give rise to the main source of conflict. Thus P3 was repelled as the county council backed down over its original stance that the scheme should go ahead in all possible parishes in the county.

Previously the Cotswolds Wardens had been able to operate within their own framework of practice, which had developed over time. This was a framework that was not necessarily concerned with strictly marking out where rights and responsibilities lay in respect to the system of rights of way. Instead the wardens were interpreting between local relations of power and national authority. This was assisted by the lack of resources that had historically accompanied countryside management and rights of way in particular.

Such practices are reflective of the local standing conditions being at variance with the legal distribution of rights and responsibilities and are illustrative of how local citizenships may evolve. The wardens claimed competence using the discourse of stewardship; that their experience and length of operation legitimated their practices. This was used to undermine the P3 scheme and any potential volunteers within the AONB – again an example where benevolence on the part of volunteers and its status as 'gift' can be used as a threat (see Parker and Ravenscroft, 1999).

Local resistance to P3

The focus of this second part of the narrative is on the reaction to the P3 scheme in the locality of Parish 'A' in north Gloucestershire. It is located within commutable distance from several large towns in the region, yet the area is dominated, in terms of land take, by small farms. Historically this spatial dynamic was linked to the reliance on the agricultural sector. These were also related to the materiality that had been underpinned by the productivist regime. The situation of Parish 'A' may in that regard be indicative of wider cultural change and conflict to be uncovered elsewhere, owing partly to factors of inmigration and increasing leisure use in rural areas (see Clark *et al.*, 1994; Boyle and Halfacree, 1998). This example illustrates how the locality has developed standing conditions that form part of the cultural arbitrary; arguably, such conditions had little cause to change rapidly, or in terms of balance of interest, since say the Scott Report of 1942. Those conditions historically left legal rights of way rights and responsibilities unregarded and a different kind of material vacuum, liminal space or 'white map' territory to evolve in terms of spatial practice (see Crouch, 1999).

In the area of Parish 'A' there had been little history of active management of the rights of way network and a number of paths were blocked or unused. In this regard, rights and responsibilities as designated at the national level were not routinised, therefore local culture operated 'around' those legal prescriptions. A small group of volunteers headed by a semi-retired parishioner entered Parish 'A' into the P3 scheme. The parish council (dominated by farmers) initially welcomed this act but after a while there was consternation from some sections of the community. As mentioned, traditionally many paths had been unmarked and blocked in the locality and part of the planned works under the scheme involved placing signposts and waymarks in the network of paths within Parish 'A' and adjoining parishes.

When the scheme actually got under way, problems arose as people realised what it meant on the ground and how it affected the standing conditions. This specifically, but probably not exclusively, related to the disturbance of previously unmarked or unused rights of way, thereby clearly indicating the contrast between legal rights and local practice. The standing conditions here had allowed for wide-scale disregard for state-set rights and responsibilities. Such disregard was partly of course because of a lack of demand for use, but partly for want of signifiers and a conducive cultural environment. In the same parish, alterations in rights of way management were attributed to particular individuals and rippled outwards, being expressed, for example, in terms of how the village hall was used and who took part in organising events (see Parker, 1997).

Resistance can be detected in both physical and discursive forms as well as being discernible in detail – part of a wider discourse analysis. A clear example of physical or material resistance is encapsulated by the reaction to the new signs that the P3 scheme funded for rights of way in the area. As soon as new rights of way signs were erected in the area, black bags were put over many of them and new waymarkers were levered from posts and gates in the area. On several occasions the P3 co-ordinator was asked to attend meetings of the parish council without being told why and was promptly rounded on. He describes one such incident:

'I was set up because I walked down there; they [the parish council] would invite me. As soon as I got there, I'd see all the farmers in their best suits were coming from all around – they'd been alerted you see . . . after an hour and a half of being bashed around, metaphorically, I would come home absolutely fuming.'

(P3 co-ordinator, Parish 'A')

This period of conflict continued for several months and culminated in a meeting that involved the local MP, the Ramblers' Association and officers from the county council, as well as local people. This placed the conflict on a higher level, bringing in agents from outside the social field, in similar fashion to the Wye example where 'outside authority' was mobilised in order to resolve the dispute. Regardless of 'sympathies' for either side, the 'national' agents at that point had to attempt to demand the performance of national (legally defined) conditions of citizenship and enforce the state-set legal rights and responsibilities. Here we see distant actors being pulled in to override local standing conditions once the conflict was taken to this level. As in the case of the Cotswolds Wardens, the threat of dividing 'rural community' and polarising opinion was invoked:

'One of the farmers said to me, in front of the local MP, that this was "the most divisive issue that he had ever known in the village", and he was born in the village. . . .

> . . . some people are so anti-footpaths for whatever reason that it has been reported back to me that they have said that "we will have nothing to do with village activities while these people [the interviewee and others] are in charge".'

(P3 co-ordinator, Parish 'A')

Eventually the rights of way in Parish 'A' and surrounding parishes were successfully opened up by the volunteers and the walks designed by the P3 group have become well used, especially during the spring and summer months. The group advertise the walks and have produced high-quality leaflets to interpret the routes – now labelled after aspects of local heritage and flora and fauna that may be found en route.

In this instance the local standing conditions were overcome, although the P3 volunteers needed to call in other authoritative actors to support them. The victory may have been pyrrhic: the experience has jaundiced the P3 co-ordinator's view of the village and of voluntary work more generally. He claims to rely more on other networks external to village life and now takes part in no village activity except the rights of way work that he and others had initiated. He no longer feels that he is 'part of the community'. Thus he has paid a price for challenging the standing conditions. His actions represented the national scale of governance, which was seen by the parish council to be imposing rules on an established order, or skein of relations, that had operated informally and was based on the local scale, rather than on 'national' knowledges. The knowledges that were in force prior to the P3 scheme were more about relations than about place or about the even performance of citizen rights and obligations.

As Goodwin (1999) notes, increasing diversity in the countryside points towards change and contingency, thus creating new possibilities. Fluid space is still punctuated by national and international levels of regulation and structure even if they are contested and appropriated locally or by certain sociations (see Shields, 1991; Parker, 1999a). This increasing diversity, coupled with agricultural crisis and other associated socio-economic problems, means that country-dwellers, and other powerful interests who look to 'steward' the countryside, are being pushed further towards accepting a detraditionalisation in their terms. In terms of a Bourdieuian reading this is a reworking of the field of power, and for Foucault a reconstitution of power relations (see, for example, Tait and Campbell, 2000).

'Active' citizenship: status, identity and activity revisited

In Chapter 5 the Local Heritage Initiative (LHI) was discussed in relation to the Diggers example. It also has resonance here. The heritage baton has been grasped through schemes such as the LHI, in similar fashion to other conservation schemes of the 1990s – P3, Rural Action and Parish Maps, for example – that have invoked or anticipated active citizenship (Goodwin,

1999; Martin, 1995; Crouch and Matless, 1996). Such policies encourage local people to engage in designing and representing aspects of their locality. The LHI actively invites the use of 'old time' or history/heritage whereas P3 implies the embrace of new time legitimated by a contested but authoritative spatialisation – that is, the Definitive Map and associated rights and responsibilities (Ravenscroft, 1998). In this way the reading of citizenship here is multiple, relating to activities of challenge that can be ascribed to a fluid, late-modern or 'postmodern' politics where diverse cultural resources are drawn upon as part of preferred 'good', 'active' citizenships.

On a more prosaic level, the P3 scheme has also served the purpose of illustrating how the meanings attributed to key words such as citizenship, participation and empowerment can be used to underpin what appears, in the case of P3, to amount to a work programme and grant-aid opportunity for a quango and for local authorities. Both of these were faced with the difficult task of fulfilling statutory responsibilities in the area of countryside management. Appropriation of citizenship can be seen as a discursive ploy on the part of government, but with interesting consequences as all parties in the interactions discussed attempted to use, rework and resist the P3 scheme and the rhetoric of citizenship and 'rights'.

P3 was devised under the auspices of the Major government, and arguably little has changed in terms of the Blair government's approach. Good citizenship in that model appears to be one that relies on careful tutoring to ensure conformity to established practice – in short, citizenship as doing what one is told, where one is told and when one is allowed. The reciprocation is to do things how the state expects, presumably by participating in the P3 scheme or the LHI, by voting, and being generally law-abiding. This is very much the approach of organisations such as the Institute for Citizenship, which has as its primary focus the 'educating' of people to be 'good' and 'active' citizens. It is argued that citizenship may require information and knowledge, but to suppose that it can be 'taught' by the state appears narrow-sighted, extremely infeasible as well as particularist, partial and exclusive.

In contrast to the government view stand the resistance and unforeseen outcomes to be had even in the P3 scheme; as the countryside changes, culture changes, incrementally and unevenly. Such citizenships are learnt unevenly and from multiple sources rather than being taught uniformly. The conflicts uncovered between landowners and P3 volunteers and between P3 organisers and pre-existing voluntary groups pay testament to this. In both examples, contestation took place over resources, and the actors involved ranged from local individuals through to national-level policy-makers and media, while all seemed to involve active citizens and activity in different forms. The disputes and conflicts challenged both national policy and local interpretation. Once local standing conditions were challenged through the P3 scheme, the Cotswolds Wardens resisted changes to the cultural arbitrary. Their tactic was to threaten exit – a symbolic and practical problem in terms of the political and the managerial relations in Gloucestershire. It could be said that the wardens had become intermediaries of local power relations while the P3 scheme volunteers and officers were intermediaries of state power.

Thus the P3 scheme provides an example where separate groups hold capillary power, where competing legitimacies are gained from locally specific power and from the state and where those competing definitions do not allow for mutual coexistence. The achieving of distinction through their relations and the resources that they can bring to bear in maintaining distinctiveness was challenged (see Strathern, 1999). Such distinctiveness must also be authoritative or bear legitimate distinctiveness. In the case of the Cotswolds Wardens, such a distinction was achieved by the invocation of notions of cultural/technical competence and the wider 'discourse of stewardship' in order to justify their practices and to critique the impact of P3 (Parker and Ravenscroft, 1999; Woods, 1997a; Cloke and Little, 1997).

While the P3 policy of itself would not usually be regarded as 'important' (it was not large scale, did not use large amounts of public money and was not widely perceived as being 'political'), it is suggested here that it could be considered as such. It is microcosmic, and many other (rural) schemes and activities could similarly be treated as being materially involved in the political, in the sense that they prompt change, even if that change is temporary, marginal, uneven or localised. They also spark conflict. While Goodwin (1999) has viewed P3 as being part of the wider drive towards local participation in 'conservation', here a rather different reading of P3 has been set out. The scheme was essentially seen as a means to achieve a pre-set objective: part perhaps of a 'totalising vision' that Goodwin has depicted. The Countryside Commission had a vision, a problematisation of rights of way and amenity use of the countryside. This provided a means of entry for willing volunteers to engage with aspects of their environment – not only an environment of objects, but also an environment (or field/relations) of power. In this sense, P3 became part of a process of territorial contestation.

'Active' citizenship and the state

The citizenship rhetoric employed by the state did provide a legitimation for some P3 participants to challenge local practices, despite the rather narrow and limited notion of active citizenship through sponsored practice promoted by the scheme. It should also be highlighted that all the examples considered here (and in Chapter 5) relate to individuals who have moved themselves to engage proactively or reactively with 'authority'. Conversely, the absent majority were not involved in these interplays or implementation of policy. The issue of apathy and lack of engagement, as assessed by normative conceptualisations of citizenship, is rejoined and debated in Chapters 7 and 8.

When one is assessing a scheme such as P3, the activity engendered should be considered as well as the discursive positioning of the actors involved or touched by the scheme. The consideration of activity as symbolic and powerful in its own right, as suggested here, needs further exploration. It is asserted that citizenship might more appropriately be regarded as processual and relational. Citizenship can be seen as being part of individual status, identity and as a consequence of activity undertaken or not undertaken. For example, one might be a good national citizen and a bad local citizen, particularly if assessed using state-defined assessment criteria of citizenship. Similarly, one may simply 'engage' with some issues and aspects of a citizenship code some of the time, or only under particular circumstances. It is also persuasive that citizenship cannot be measured comprehensively or objectively. Indeed, it is at all times disparate, subjective, partial and a personal, if not cumulative, process.

Citizenship as activity takes place within wider social and political contexts, and the conceptual model of citizenship being applied here is processual. The critique in Chapter 2 of the Marshallian model of citizenship as evolutionary and linear can be applied to Luhmann's (1982) view of activity as 'apprenticeship' – particularly any implication of a static reading of 'competence'. Fluid space implies fluid practice, necessitating a learning or 'becoming' in the technical sense, but also a need to learn the implications of practice, or indeed the development of a reflexivity that is concerned with the need *for* change. In this sense, the becoming of citizenship may be more about nurturing the ability to be reactive rather than (or as well as) proactive.

Active citizenship may be promoted and government or the state may envisage particular objectives more widely, but such activity cannot necessarily be controlled and the outcomes or diffuse repercussions of practice cannot be totally anticipated (or even be immediately identified). While recent active citizenship projects imply a narrow (e.g. citizenship as voting) and/or apolitical participation (as envisaged in the P3 scheme), it is not so: people alter place and rights through their actions. However, on reflection, it does seem a pity, on at least two counts, that democratic governments in the UK cannot contemplate deeper citizenship. First, they seem not to understand widened accounts of citizenship, and second, in drawing tight and limited lines around 'active' citizenship they still misunderstand how culture and practice are engines for change and resistance – witting and unwitting.

Activity and engagement (or non-engagement) is also part of individual positionings within the social field – that is, it is constitutive of claims to status and identity (linked in some sense at least to space, territoriality and topophilia (see Tuan, 1974; Shields, 1991; Cresswell, 1996). Crouch, for example, has long argued that the everyday, the mundane is a key aspect of cultural, political geographies (see Crouch, 1999) and presumably of citizenship. Such an analysis begs certain important questions. For example, what is the everyday? And is citizenship to be regarded as something extraordinary, or to be set apart from the everyday? Certainly the state portrays citizenship as 'extra-ordinary'. Can it be detected in what people *actually do* rather than what they 'should' do? If the latter is the case, then this ties in with an important element of each of the examples thus far: that citizenship may be more about challenging and resisting than about assisting the state and elected government in implementing, or at least not obstructing, mandated policies or 'legal' activity.

The contested definitions in this chapter have tended to relate to priorities and rights/responsibilities with regard to access in the countryside. It is likely that actors attempt to persuade others that particular approaches and issues are more or less important than others, and therefore practice and regulation are cast and recast. It is argued that there are increasing potentials for conflict within a changing, fluid countryside, particularly given the way in which social relations at the micro level are mediated and where the countryside is contested by increasingly diverse interests and individuals. When compared with the Wye example the issue of routinisation appears to be a common factor in the P3 and Wye examples. They show contrasting scenarios where legal rights were not exercised, and where attempts were made to exercise such rights, standing conditions and local practice were challenged.

In such cases, acute conflicts arise between different interests and scales. Such change involves standing conditions becoming liable to change and implies that the rules of the social field can be misread, ignored or created anew by various actors within that field. Marsden *et al.* (1993: 29–30) argue that this is partly

> because it adjusts the social basis of entry from ones of customary rights to those based upon economic power . . . also the social processes underlying these changes in the definition and trading of goods and services, and their realization through political action at the local, national and international scales.

Where standing conditions are sidelined by some, it may become easier for others, perhaps particularly local people and younger generations, to ignore them and play a part in further altering those conditions. Thus each challenge and alteration in practice and consequent rights distribution/ sanction plays a part in socio-economic change and a respatialising/ reterritorialisation of the countryside. Hence, one lesson that an apparently innocuous scheme illustrates is how attempts to change practice and rights and responsibilities that are developing unevenly (and sometimes illegally) are difficult for governments to halt or reverse and that standard conceptions of 'good' citizenship activity are received differentially and enacted differentially. In many other aspects of rural policy and practice, difference flourishes and is challenged (or challenges) only when in conflict with an alternative, competing definition or claim that is marshalled with the resources necessary to sustain it.

In Chapter 7 another facet of citizenship and active citizenship is set out. How citizens *qua* consumers engage with governments and the corporate world is examined and the role of the consumer-citizen is related to past actions and the future in the countryside. It is also shown how single-issue politics, protest, consumption and consumer-citizenship may be assessed and linked to the countryside. Then Chapter 8 incorporates the key aspects detailed here in a synthesis that reviews empowerment, post-national citizenship and the role of different scales in organising and deciding policy. Chapter 8 also synthesises the different aspects of the politics and practice of citizenship examined in the case studies.

7 Citizenship and the countryside as consumer space

Introduction

While this text centres on the rural, there is undoubtedly overlap with and yet a semi-conscious division between environment studies and rural studies. The rural and the environmental are conflated at times, with perhaps the rural being seen as a sub-set of the environment and therefore conceptualised as a resource to be conserved, preserved or otherwise 'protected'. This chapter is concerned with the 'rural environment' and rural environmental citizenship as introduced earlier in the text. This field of view is used as a means of approaching the issue of consumption and consumers in terms of the rural. Here other examples of citizenship as activity (linked to identity politics) are illustrated, showing how attempts to define legitimate action are being resisted and restructured by interests in the rural environment. The wider arena of environmental action and the market is considered where the space being contested is at once 'rural', environmental and also contested in terms of rights (of consumption and contractual entitlements) and activity.

That statement begs certain questions: how is 'citizenship as activity' delimited? Are the actions of those participating in the Countryside March of 1998 'active' citizens, for example? Perhaps in a similar way to those engaged in actions against road-building projects? Part of the aim here is to focus in more detail on fragments of such a widened conceptualisation of citizenship in the context of the countryside and to see how particular countryside spaces and practices are contested – in particular, how markets and consumers engage in shaping the countryside both in the mercantile sense and as amenity-users. This aspect of politics is receiving increasing attention (see, for example, Klein, 2000; Monbiot, 2000). This chapter therefore examines both the 'single-issue' politics of the rural in the 1990s and overlapping 'environmental' politics in the same period. It does so by means of two main examples: the Countryside March and the Newbury bypass protests. Mention is also made of a range of other actions and policies that may be considered relevant.

In both Chapters 5 and 6 examples were detailed of protest and activity that involved alternative forms of agency and resistance towards attempts by national government to activate citizens. The P3 scheme illustrated the micro-politics involved in the implementation and resistance to policy involving citizen action. The Diggers example in Chapter 5 showed how such sites have been selected on the basis of their 'heritage', and there is definite connection in the examples given in the present chapter. The framing of citizenship and the constraining of (legal) agency affects the way that people behave or feel about political and other lifeworld elements. Sack (1993: 57) argues that 'we should make choices that change structures so we can have "real" alternatives and, thus, real freedom'. While this statement is important, and the P3 example goes some way to illustrate the point, it can be challenged. The environment in which change and agency/practice are nurtured is an important aspect of citizenship 'protection' on the part of the state and other powerful actors.

In previous chapters it has been argued that conceptualisations of citizenship have been narrow and legalistic, and that such definitions have lagged behind cultural theory and developments in global economy/society (see, however, Isin and Wood, 1999; Urry, 2000). It has been asserted that citizenship might usefully be viewed more fluidly and widely, giving consideration to how people make citizenship as well as how the state develops it. Similarly, there has been little written about how state-defined 'active' citizenship squares with other individual or sociate actions that have not been defined by the state as being within the ambit of citizenship. As a consequence, the potential scope and nature of the 'activity' in which the citizen has 'legitimate' interest have received too little attention. We need to consider how practices and flows of resources or capital, widely defined (see Healey, 1997; Bourdieu, 1990), impact on the rural environment and rural imagination in both physical and psychological terms.

Urry has argued that there is a struggle in global politics between identity politics of the 'jihad' and the consumerist model of 'McDonaldisation' (2000: 209). On the domestic and local levels, identity and single-issue politics are similarly ascendant, seeking to appropriate consumerist or market-based tactics as a method of pursuing rights-claims and other political objectives. In essence these are attempts at using dominant economic modes to pursue political and moral ends. The rise of consumption-based tactics and expressions of 'consumer-citizenship' relate to debates as to whether citizenship in a consumer society can be bought and whether some rights are becoming consumer rights available to those with the means or capital (in its different guises) to 'buy' or realise them. If rights are privatised, do the actions and practices involved thereof cease to be parts of citizenship? It is argued not, and even if the answer were only a partial 'yes', it still shows deficiencies in arguments and evaluations of citizenship based on state-defined citizenship. It is also logical to examine how citizens react and manipulate consumption and how producers and structuring agents themselves react to consumer-citizen pressure. Within advanced or late-capitalist societies this situation is often presented as preferable to alternative methods of receiving or exercising civil and political rights. This is particularly so considering the role and influence that the corporate world has on both formal politics and culture into the new millennium.

In Chapter 2, and subsequently, citizenship theory was examined and discussed in relation to culture and society, with some emphasis on the re-evaluation of citizenship under postmodernity or late modernity, and an appreciation in a more expansive, if partial, manner was attempted. The examples drawn upon thus far have been illustrating how citizenship and expressions of citizenship are diverse. They are reflections and examples of resistance to state actions, local actions and contestations over extant rights and responsibilities. Action groups or interest groups may be viewed as representing new coalitions or networks in post-national society; they represent a layer of governance and present a level of meaning for those subscribing to the 'interest' or aspiration that binds the group. Such aspatial, place-specific or dispersed communities operate using information and computer technology as well as other technologies, other media and the market, together with cultural capital and normative, legal or processual engagement with antagonists; indeed, a variety of means and tactics are used (see, for example, Kousis, 1998; Parker, 1999c).

Rural politics and politics more generally have been undergoing significant change over the past thirty years. Politics has altered in parallel with technological developments and other social change that have made politics more disseminated, visual and mediated (see Hay, 1996; Anderson, 1997; Routledge, 1997; Urry, 2000). This chapter has a focus on the 1990s. In that time frame one of the most significant developments in world culture was the rising and changing influence of the media. Newspapers, and television in particular, have impacted on the performance and packaging of politics. In the 1970s particularly the media were viewed more as a means of disseminating ruling-class ideology (see Cohen, 1972). Since then it has become apparent that other interests have learnt how to manipulate the media and how to provide 'stories' that may be recounted by various media forms including mass readership newspapers and television channels. As interests become more media aware and recognise how messages and ideas can be communicated powerfully through TV and newspapers, they plan political actions or photo-opportunities with that media-tion of political action in mind.

The media have become more susceptible to being used by a variety of interests as a means of amplifying claims and ideologies. In consumer society the media provide the spectacles for consumption. It is also the case that advances in communication technologies have made it possible to pass information and get information more readily as part of a process of engaging in politics or in the politicisation of particular issues. Numerous political campaigns of different kinds have relied on the media to disseminate messages. Quite literally, the media have become the single most important method of political communication between groups. More widely, the media industry is being used relentlessly by commercial interests to advertise and sell products and services as part of what has been termed the culture industry (Adorno, 1990; Molotch, 1999). In Baudrillard's terms, both forms are part of a sign economy in which politics becomes aestheticised as another form of mediated consumption (Baudrillard, 1988; Miles, 1998; Routledge, 1997). In this sense the connection between the media, politics and consumption as applied to rural Britain is made.

This section also examines how the countryside has been represented and contested as a homogeneous or somehow separate entity by certain communities of interest in recent years. This representation of the countryside is emphasised through a review of the mobilisation of such interests by the Countryside Alliance, and how those attempts to present the countryside were received and resisted by other communities of interest. This also implicates how traditional discourses in an era of mobile and post-national citizenships (Urry, 2000) are important elements in such struggles. Finally, a discussion of the countryside as consumer space links the two examples used in this chapter, detailing the protests over the Newbury Bypass in the mid-1990s and the Countryside March in 1998. Both are indicative of ongoing wider struggles to mediate, resist and structure change in rural economy and society.

'Postmodern' politics, media-tion and communities of interest

It has been argued that the physical environment is socially constructed, and Urry (1995), building on such an assertion, views the rural in similar fashion: the countryside is both socially constructed and moulded by relations of consumption as well as production. While rural studies has traditionally focused on actors with a tangible 'stake' in the countryside as a productive space, such as farmers, landowners and other rural-dwellers or workers, it is more inclusive and justifiable to consider all the groups and interests who are touched and concerned by the countryside. In essence an argument can be made that all groups are in some way touched and concerned. This is in contradiction to the rhetoric of traditional rural interests, who tend to claim unique knowledge, skills or 'rights' in connection with the rural – a category that such interests attempt to reserve definition of. As explored later, this position involves considering the actions and opinions of a wide range of interest groups and economic agents, something that is sketched out but cannot be covered in detail in the confines of this volume.

In Chapter 3 I pointed out that under a condition of postmodernity and in an era of rapid communications, concomitant with other features of globalisation, the way that people express their views and take part in politics is varied and often intermittent, or unstructured. People look at alternative strategies or lines of resistance that are available. In Urry's (2000) terms, such actions are part of mobile and post-national citizenships, where simple notions of territorial and state-defined practice are less relevant. In such circumstances it is possible that political engagement is less and less likely to be expressed along class or party political lines and may relate to single issues or issues that are distant in terms of space (and perhaps time). Those issues are also likely to be promoted as being important by particular communities of interest as and when they coalesce (perhaps coalescing around a particular issue). This process may be viewed as a consequence of both the information age and globalisation generally, and also of the so-called democratic deficit whereby people seek to engage with issues that directly affect or concern them regardless of 'process', legal jurisdiction or mode of expression. Such tactics have tended to arise regardless of the democratic model operated within a territory. The democratic deficit might of itself be partly explained by processes of deterritorialisation and through blaming national governments when they appear unable or unwilling to incorporate or act upon diverse interest-claims, or, as discussed later, to develop mobile or fluid structures that can cope with social and cultural fluidity.

Postmodern political action may be fragmentary, be of a short duration or be consciously organised to generate media attention – thus relying on 'media-tion' (Routledge, 1997; Anderson, 1997). Alternatively, as with recent anti-paedophile demonstrations in England, the media can lead such actions. In this case the Sunday newspaper the *News of the World* publicised the names of alleged offenders, thus providing the information to reignite a moral, local and locational panic that led to a series of riots and unfortunate cases of mistaken identity (*Guardian*, 2000b).

Practices may be informed from a diversity of sources and views aired that are derived or based on image and imagination that involves the visualisation of 'collectively held social meanings among the occupants and users of a place' (Stokols and Schumaker, 1981: 446), notwithstanding sociations who adopt such meanings. In that sense, people do not have to have direct experience of a place setting to have a sense of place (Tuan, 1974) or adopt claims to rights, but they may become part of a 'place system' (Bonnes and Secchiaroli, 1995) when their (distant) actions affect places. Following this line through, it may be argued that consumers can politicise their actions in response to campaigns or through personal experiences in order to express views in terms of other practices and impacts on place/environment. Those actions can dictate a range of *de facto* rights and responsibilities. Consumers exercise rights and demand responsibilities from corporations where the state will not guarantee or enforce, or has not guaranteed or enforced, certain entitlements and obligations. This transmutation of roles effectively means that group or societal value statements can be expressed through the market and contra to the views of other market players, governments or 'home' states.

It is also the case that groups use action spaces in order to pursue claims and to engage in politics. These spaces are deliberately selected on the basis of their symbolic value in order to intensify both message and meaning (association of place to issue) and the message's media-tion (Routledge, 1997). It has also been discussed how the topic of citizenship and the countryside is not something that can be justifiably assessed only in terms of people living and working in the countryside. It is also about how interests impact upon the countryside from afar or indirectly. For example, anti-hunting may equal anti-traditional, or elite rural, or pro-animal rights (which is indicative of the multipleness of what are apparently 'single issues'; see Macnaghten and Urry, 1998). It is mentioned here how rural space has been contested within urban space (e.g. the Countryside March and Rally); and vice versa, how despoliation or 'urbanisation' is contested on eco/environmental grounds in the rural (e.g. the Newbury road protest) is similarly discussed (see Urry, 2000).

Critical linkages are made between the concept of citizenship (including 'active' citizenship and cultural citizenship), consumerism, protest, identity and the market. Once again, as explored in Chapter 6, leisure use of the countryside and conflict can make for a useful topic area with which to examine issues of power and the construction of appropriate activity and rights/obligations in the countryside. Two examples are used. First, the countryside protests organised by the Countryside Alliance in 1997 and 1998 illustrate how previously dominant rural interests have attempted to resist threats to their leisure practices; they are linked perhaps to symbolic power struggles and changing socio-cultural relations in the countryside. Second, we shall see that consumer-citizenship may influence rural, and related environmental, affairs by prompting 'action through buying' (see Parker, 1999c). A descriptive account of actions aimed against road-building at Newbury in 1995–96 makes the point. The examples detail the way that postmodern politics is crowded with multiple (but not necessarily inclusive) voices and claims, and show that resistance to the market and to the actions of business and government towards the rural environment is expressed using both direct action and market-based tactics.

Consumer-citizenship

The differences between the *consumer* and the *citizen* have been focused upon by many writers (cf. Gyford, 1991; Lowe *et al.*, 1993; Urry, 1995, 2000). Citizens exercise rights through the political system and consumers react to the market mechanism. Featherstone (1991) and Urry (1995) note how consumption through material goods or other consumption practices provide a channel of communication for consumers to other members of society and to those in positions of power over them. It is evident that political action is increasingly being taken through consumption practices. Urry (1995) argues that citizenship as political status and that of consumer as economic agent are merging and it is likely that interactions between citizens and the business world will intensify, with the roles of consumer and citizen continuing to overlap. Frequent calls to boycott certain products or companies are likely to continue – some with more success than others. The approach is largely issue or interest based and has already featured in a number of environmental protests, thus having clear implications for the regulation and management of the rural environment.

In terms of more mainstream consumer protest, the United States witnessed the first use of large-scale consumer protest tactics during the 1960s and 1970s. Ralph Nader in particular promoted the use of a consumer-citizenship approach in the United States, with some success (see Nader, 1990; Harbrecht, 1989; Mintzberg, 1983). In the UK, environmental groups have been using and publicising market-based tactics such as No Shopping Days, or for example Christian Aid's Change the Rules campaign (which encouraged consumers to register protest directly at the supermarket). More generally, a multiplicity of interests have begun the boycotting of particular shops or companies because of perceived poor 'corporate citizenship'. These actions can be read as an important part of the shift towards 'lifestyle politics'. Other European countries have witnessed similar tactics; at the time of writing, French lorry and tractor drivers were blockading the roads, seaports and airports over high fuel prices in an effort to persuade their government and OPEC to reduce the price of crude oil. The (mediated) protests also spread to other EU countries, including the UK, with the initial emphasis on registering protest at national taxes (and therefore in 'performing' home states) but indicating how global economics determine such affairs (*Guardian*, 2000b).

The consumer-citizen is a construction defined and exercised politically. Therefore the construction and applicability of notions of citizenship or of the consumer are at issue; where, when, and why individuals are called upon to act as consumers or citizens in particular situations and how those roles in particular situations are justified or legitimated are important. Urry (1995: 165) makes the connection between consumerism and citizenship forcefully, noting that 'people are increasingly citizens through their ability to purchase goods and services – citizenship is more a matter of consumption than of political rights and duties'. It is argued that the dynamics of consumption itself are increasingly politicised. There are already numerous situations whereby UK citizens are encouraged to be 'consumers' of public services, and this relationship is being turned around such that consumers of private services look increasingly to be able to exert political rights of citizenship over those providers. Citizens are seeking to be involved in governance or to express governmentality (see Simons, 1995: 36) where it is perceived that the state is inactive or unwilling to intervene. The 'consumer-citizen' enacts such expressions of politicised consumption.

Another reaction to such consumer-based developments on the part of food producers (and apart from the high-profile actions of organisations such as the Countryside Alliance) is shifts in producer–consumer relations – for example, through farmers' markets (see Holloway and Kneafsey, 2000;

Latacz-Lohmann and Laughton, 2000) and in terms of demands for organic and GM-free foods, partly fanned by scares over BSE and other health-related or environmental issues such as salmonella in eggs and wider animal welfare concerns. Elsewhere, including in the UK (and directed at various scales within and beyond), there have been a range of well-publicised consumer protests during the 1990s, for example against Nestlé (concerning providing baby-milk to developing nations), Barclays Bank (relating to its association with South Africa and apartheid), Shell (in connection with the Brent Spar controversy and pollution in Nigeria) and against the French state (via French produce) over its nuclear weapons-testing programme. In each case the boycotting of products or services en masse was the mode urged in order to express opposition to actions and outcomes. Thus the use of such tactics affects localities across the globe and may be motivated for a variety of reasons. There are numerous other types of action where the citizen as consumer exercises political views through the market. One example is the strategic buying of shares in companies in order to protest against company actions, thus illustrating the citizen as shareholder-protester (as discussed later – another example of buying private rights). Another is the outright buying up of land in order to 'protect' it or to create a piece-meal land ownership pattern that can frustrate new development (see Marriott, 1996).

The development of new commodity relations restricts, but provides opportunity for, resistance and engagement by the consumer as protester, in particular where the main object of engagement with the market is not to profit in monetary terms, but to regulate the actions of the company or of other interests that aim to exploit or develop a particular resource. As discussed earlier, consumer-citizenship represents an element of postmodern governmentality in an era of glocalisation, (re)territorialisation and detemporalisation. In reverse, it is worth noting that the role of consumers/citizens in and of the countryside increasingly involves the claiming of consumption rights (or leisure rights) in the countryside (Bauman, 1998; Parker and Ravenscroft, 1999; Macnaghten and Urry, 1998). This aesthetic consumption role also means that public opinion about the countryside tends to favour its preservation (but while development for leisure and associated use may be justifiable, most other forms are not).

'Listen to us': the voice of the countryside

Britain's Labour Party has never been the party of preference in the shires or of farmers and other traditional or landed interests. Most Labour support has traditionally been drawn from urban and industrial areas. Labour is consequently viewed, or portrayed, as an urban party that lacks 'understanding' or experience of rural affairs. Even before Labour were elected in May 1997 the various field sports communities and other rural business interests were preparing for that likely eventuality, with the Countryside

Movement consequently coming into being in 1995 (Countryside Movement, 1995; Keeble, 1995). This was essentially a movement funded and supported by rural business interests. Funders included gun-makers and large landowners such as the Duke of Westminster, even though it claimed to represent the countryside more widely and was fronted by David Steel (now Lord Steel), previously the leader of the UK Liberal Party and considered to be a centre-ground politician (Welsh, 1997).

By 1997 British agriculture was apparently in a desperate state, being hit by successive confidence-denting problems such as the BSE crisis and suffering from low prices for agricultural products (Fisher, 2000; Thirsk, 1999; Public Accounts Committee, 1999). The situation was being described by farmers as desperate, with many smaller or less profitable farms closing or being amalgamated with larger agri-businesses (MAFF, 2000; Hart, 2000). This general feeling of mistrust and anxiety was tapped into by the Alliance, which attacked Labour early in its administration. The new government was not seen to be acting quickly on rural issues or as having knowledge or understanding of the countryside (even though many of the issues facing the countryside and rural economy had been apparent well before the 1997 election).

The example used below is the recent protest made by a combination of communities of interest that centred around the Countryside March of 1998 (see Woods, 1998) and an earlier similar, though smaller-scale, Countryside 'Rally' organised by the British Field Sports Society (BFSS) in the summer of 1997. Both took place ostensibly to counter any threat of a hunting ban. By 1997 the Countryside Movement had joined forces with the BFSS and others and re-emerged as a new organisation, the Countryside Alliance. The Alliance has used numerous slogans and 'straplines' and has been well funded and extremely well organised in terms of public and media relations. It states that its aim has been to 'champion the countryside, country sports and the rural way of life' (Countryside Alliance, 2000).

The BFSS had been planning a demonstration to air its views on the threat of abolition or emasculation of various field sports. This was to take place in July 1997 only two months after a new government would take office. The event went ahead on 10 July under the Alliance banner instead. The timing would also coincide with the earliest opportunity of an anti-hunting bill to go before Parliament. The central issue that concerned both the BFSS and the Countryside Movement (and its prime mover the Countryside Business Group; see BBC TV, 1999; Welsh, 1997) was that of hunting and likely attempts to ban it. The threat of a hunting ban did materialise in July 1997, as a private member's bill promoted by Labour MP Mike Foster was presented to the Commons. The rally attracted around 120,000 people, and while presented as a broad platform for a range of rural issues it was a scarcely concealed pro-hunting demonstration (Monbiot, 1999; Woods, 1998).

Later in the year, at its second reading in November 1997, the Foster bill proceeded with a 411 to 151 majority vote that gave anti-hunt supporters encouragement that the government would find time for the bill in its programme. Meanwhile, the Countryside Alliance had been planning another protest, this time to include a march in central London to be held on 1 March 1998. The plan was to get as many people as possible on the streets to demonstrate dissatisfaction with the state of the rural economy. The march eventually assembled around 250,000 people and the event passed off peaceably, with no (physical) conflict with rival anti-hunt demonstrators. It gained wide press coverage, and many national politicians from all parties attended the event.

Press coverage of the event recorded how diverse the messages being promulgated were on that day (Engel, 1998). The Countryside March claimed to be the 'voice of the countryside', but it was not supported by some rural interest groups, notably the National Trust and the Council for the Protection of Rural England (CPRE). Indeed, a whole range of groups and interests were either excluded or refused to participate. The march was promoted on the basis of being 'for the countryside' – a bold claim echoing outmoded attempts to see the countryside as homogeneous. The Alliance campaign may be read as an attempt to freeze, singularise and portray the countryside as a reflection of the Alliance's own image. The Alliance represents particular interests that identify and demand particular ruralities and particular practices to survive – further than this, for the state to support those practices legally and discursively. Such ruralities include fox-hunting, grouse-shooting and other field sports.

The overall message presented by the Alliance has been that rural interests should be considered more urgently and more seriously in government thinking. Within seemingly reasonable policies for the countryside, the Alliance places particular cultural practices, as indicated, as central. Some other interest groups were latterly invited to take part in the march. The Ramblers' Association and the angling interest successfully bandwagoned the march and helped cause what has been viewed as a 'bloody coup' within the Countryside Alliance as a result (BBC TV, 1999). Unpredictably, but perhaps more representatively, the march helped provide a platform for a cacophony of often conflicting voices, as diverse interests used the opportunity to attend the event and make their own sectional point about various rural issues (Engel, 1998).

The march organisers also encouraged the wearing of particular badges of rural identity. Such symbolism included traditional dress of hunting uniforms, tweeds and other clothing such as Barbour jackets and cloth caps. Other 'props', such as hunting horns, were present to reinforce a particular imageability of the countryside and as part of a discourse of heritage, playing on a popular perception of the countryside as a national 'treasure' full of hardworking people providing essential products for all. The nationalism invoked by the Countryside Alliance in its attempts to mobilise others appeared at odds with its rhetoric. One of the slogans of the march was 'Stop the Urban Jackboot', even though part of the point of the march was supposedly to enrol widespread support for rural interests. Monbiot (1999: 92) detects that the selection of an urban enemy may be misdirected: 'the distinction between town and country promoted by the organisers of the Countryside Rally is an artificial one, nurtured by a city-based squirearchy seeking to deflect attention away from its own exploitative practices'. Additionally it can be said that such staged political actions are primarily aimed not at the public but at small groups of power-holders and (nervous) politicians.

The Alliance appeared to have swayed government thinking on the issue of hunting even before the march proceeded, with hurried negotiations and compromises being offered on hunting just before the march (*Guardian*, 1998). However, the government did not grant time on the issue in Parliament, and the Foster bill fell. A commission of inquiry, the Burns Committee, was set up instead to look into the hunting issue. It duly reported in the summer of 2000 (Home Office, 2000; Norton, 2000; Burns Committee, 2000). Despite its ambivalent findings, early in 2001 a bill was passed by the Commons on a free vote, indicating tacit governmental support. At the time of writing, however, it had yet to proceed through the Upper House.

These actions illustrate attempts by powerful and largely 'traditional' rural interests to create and exploit a cultural and imagined divide between the rural and the urban, in doing so positioning themselves as the legitimate 'voice' of the countryside, effectively marginalising or at least attempting to marginalise others with direct or indirect interests in the countryside and the rural environment. Woods (1998) sees this attempt as being assisted by the association of 'nation' and the national interest with the countryside. The Countryside Alliance, by constructing itself as the protector of the countryside, and attempting to invoke the rural as the 'national', presents a further example of the discourse of stewardship being used both as a blocking discourse and as a method of legitimating particular and exclusive or 'private' rights-claims (see Winter, 1996; Monbiot, 1999; Parker and Ravenscroft, 1999). Such a discourse is linked to a discourse of heritage which is also invoked and exploited by different interests, both to block and to engender customary and legal change (Countryside Alliance, 2000; Billinge, 1993).

The countryside may be dreamed or imagined in numerous ways. It is difficult to determine whether there is a dominant imagination – particularly now (see Short, 1991). It may be truer to say that there has been a dominant discourse that has underpinned those controlling rural space and economic activity. Such control is changing and highly visible contests signified by the Countryside March around the issues of hunting and countryside access are key indicators of that fact. Baroness Mallalieu, one of the prominent spokespersons for the Alliance (and a Labour peer), said at the time of the march that 'it is about freedom, the freedom of people to choose how they live their own lives. It is about tolerance of minorities and sadly those who live and work in the countryside are now a minority' (Countryside Alliance, 2000). In essence, the attempt was made by the Countryside Alliance to claim minority status on the basis of numerics. Contradictorily, attempts were also made to justify listening to the Alliance because 75 per cent of land cover is 'rural' (Hanbury-Tenison, 1997), implying that the 'minority' deserved special consideration as they effectively manage the countryside on behalf of the majority.

Another interpretation of the nationalist discourse is that traditionally powerful interests are losing power and influence. A new feeling of beleaguerment was being experienced (see Cox and Winter, 1997): they were a new minority. This provides an interesting point in that if citizenship is not evolutionary, then the people, activities and places from which rights can be stripped may create or reveal 'new' minorities in terms of power and cultural acceptability/capital. The nationalist discourse here is a mobilising and modernist one, but it is interesting to recall how, historically, the 'national' and its relation to the countryside and the land ('land fit for heroes') had been used previously as a means of galvanising the population – young males to go off and fight in the world wars, and latterly for women to occupy previously male-dominated jobs.

There has been an important definitional issue at stake within key moments such as the Countryside March and the debate about hunting: who is rural and whose concerns are of the rural? Moreover, who controls such a definition? Farmers and those involved in, or subscribing to, particular sporting or leisure activities are in a small minority of the overall population. In the past, the ability to control rural space has been effected by a coalition of interest on the part of landowners, farmers, the state and, in the main, of the consumer, with the access issue being an obvious counterpoint to the hegemonic relations sustained in the post-war period. The traditional interests that backed the Alliance wished to identify themselves as being in some sense the 'heart of the nation' and with the idea that change (at least that which they disagreed with) would harm the 'nation'. This play can be read as a defence of the historic control of the rural as productivist and effectively stewarded by those that know best – still resistant to the idea of an informed and critical (consumer) citizenry. Of itself, such a discursive tactic may be seen as part of the discourse of stewardship – a wider version that asks the population at large to agree that their imagined rural and those who guard it are guardians of the nation.

Thirsk (1999) argues that protest and crisis over British agriculture have been a feature through history. It might be argued, however, that the issue that is being highlighted in the example and associated debates in the late 1990s is not so much about agriculture. Rather, it is about culture and attempts to retain certain degrees of homogeneity – often expressed by groups such as the BFSS and the Countryside Alliance in terms of 'ways of life'. The problem is that the ways of life platformed by them are minority, often exclusionary and potentially unsustainable in a more consumer-scrutinised and consumerist, perhaps post-productivist and environmentally facing, countryside. The issue of hunting also links into issues of trespass and land rights, as hunts often make incursions onto private land without permission, potentially damaging crops and causing other inconveniences (Burns Committee, 2000).

In connecting and situating the march within changes in global economic and social change, Woods (1998b: 5) states that 'the emphasis of the rural economy shifted from primary production to the commodification of rurality itself, consumed through tourism, counterurbanisation and lifestyle shopping, trading on stylised nostalgic and sanitised representations of rural life and society'. Such shifts in practice and in demand have wrought a new rural politics, quite apart from issues of levels and modes of production. In particular, even if unevenly (Marsden, 1995), as the source of much of the economic power, the land is changing in terms of regulation, production, cultural relations and use. Closely associated to this change is a growing and vocal proportion of rural-dwellers who do not share the same values as Countryside Alliance supporters – they are members of different communities. Perhaps for the first time, the former group hold critical levels of power and influence. Essentially, the march and the evolution of the Countryside Alliance itself are closely connected to cultural change both in the rural and more widely in the UK.

As discussed in earlier chapters, citizenship has been constructed and defined exclusively, and state-constructed citizenship may also be viewed as an attempt to marginalise various methods of political action (see Turner, 1986; Parker, 1999a). There have always been opportunities and examples of subversion, resistance and appropriation. However, the state tends to lag behind innovations in political action and cannot keep pace with capital in its continuous economic restructuring. The Alliance sought to defend particular rights and attempted to do so by associating those rights with both heritage and nation. Other interests, meanwhile, view hunting as outmoded, barbaric even, and elitist (not to mention the animal rights issues carefully avoided by the Alliance).

The hunting issue illustrates how rurality has become an increasingly contested and multiple concept, and similarly the way it is managed and regulated is contested. Certain activities or benefits remain little changed while others are challenged and concessions made. This is largely as a result of socio-economic changes, as mentioned, that have been under way in the countryside. Different 'cultural communities' (Healey, 1997: 37) develop and gain influence whereby thoughts and worldviews are in competition. While the above example shows how an interest mobilised in order to resist a challenge to its rights (perhaps sharing some similarity with the River Wye example discussed in Chapter 6), the Countryside Alliance deployed its resources, in both pecuniary and social capital terms, to get 250,000 people to express or 'stage' their viewpoints about the state of rural Britain. As organiser, or network-builder, the Countryside Alliance presented that variety of sub-interests as being part of its own agenda. Its priority was to resist a hunting ban, but other issues such as proposals, now legislated upon, for a 'right to roam' were also of prime concern to the Alliance at the time (see DETR, 2000a; Shoard, 1999, 2000).

The politics of the (post)rural are flavoured by a host of issues and concerns and are being engaged by and related to a far wider and diverse polity than in the past. It is based on the conflicts between many competing interests and concerns that view the countryside as a resource for different, often incompatible, uses and providing different benefits that can be enjoyed or exploited in different ways (Macnaghten and Urry, 1998). McKay (1998), in commenting on DIY culture and single-issue politics, notes how such issues are interrelated and overlapping, and that the idea of a single issue may of itself be contestable. Perhaps it is unsurprising then that protest aiming to represent as diverse a category as the 'countryside' was notable for the range of issues being platformed.

In parallel to political regulation, the commodification of the rural has been proceeding gradually and unevenly since the enclosures, while the sign economy has also used aspects of the rural to its advantage, and associated or 'connoted' artefacts (see Baudrillard, 1998; Macnaghten and Urry, 1998). The development of 'consumer-citizenship' in various guises, whereby the expression of political rights and responsibilities is channelled through market discourses, may be viewed as an area where this has happened – partly because of the tardiness of government as well as the oppression of alternative means of engagement (Parker, 1999c). In the next section another example of rural protest in relation to politicised, commodified relations is considered, and in Chapter 8 a review of and speculation about the 'branded' commodified countryside and its relationship to 'postmodern' citizenship are included.

Citizenship, consumers and space/place

There are many methods of registering protest and of expressing 'citizenship', and conversely there are multiple methods aimed at attempting to engage citizens. These include forms of engagement using the market and playing on the role of consumer (see also Klein, 2000). The consideration of citizenship and the relationship between citizens and the market is focused upon here. The strategy of using consumer power is not a new phenomenon, even in the context of the environment, yet increasingly the market (in terms of practice) and the market discourse (in terms of posture and approach) are becoming legitimate mechanisms by which to register protest. The state increasingly relies on the individual, the consumer-citizen, to regulate private business (see Featherstone, 1991; Warde, 1994; Urry, 1995; Parker, 1999c). Consumer protest through the market or in using the discourse of the market is one mode of expressing and mobilising large numbers against particular organisations or to register a point with government.

This situation has arisen for numerous reasons: as a result of the dominance of transnational companies, the increasing awareness of ethical issues on the part of consumers, the rise of the 'Risk Society' and an overall decline in public faith in traditional politics and politicians. In the UK this process of disillusionment and destabilisation was almost certainly expedited by the New Right ideology adopted by the US and UK governments during the 1980s and early 1990s, an ideology that promoted self-help and individual choice over state, community or societal rights and duties.

As a result of this, it is not only individuals who were expected to adopt rational and regulatory behaviours. In terms of responsibilities and attitudes towards extra-juridical obligations, private business has increasingly been asked to take moral or ethical positions and take responsibility for its actions and their impacts on people, cultures and environments. The business decisions of private companies were being judged in terms of their effects on citizenship and rights at different scales, including their impacts on the environment. In the 1990s private-sector organisations, in particular larger corporations, began to realise that they had to engage with consumer protest strategies that were being adopted (New Economics Foundation, 1999; Parker, 1999c; Klein 2000). Recently, large companies have lobbied the UK government for clearer guidance on social and environmental responsibility – essentially for a framework for 'corporate citizenship' (Cowe, 1999; Tomorrow's Company, 2000). For example, the Institute of Citizenship (now the Institute for Citizenship Studies, ICS), set up in the UK following the Plant Inquiry into citizenship, has been working to encourage enhanced 'corporate citizenship' whereby businesses pay regard to their 'responsibilities' to the public, and by association the moral and ethical 'concerns' of the public (see Institute of Citizenship, 1992; Institute for Citizenship Studies, 2000; McIntosh *et al.*, 1998).

This framework relies on full information and vigilance on the part of consumers and interest groups; it also assumes that business can and will adopt 'moral' frameworks. Notwithstanding those criticisms, it has been claimed that businesses generally prefer to be told how to behave through directive regulation rather than cede competitive advantages by making individual pledges about future behaviour (Ketola, 1997). In short, businesses do not like uncertainty and are risk averse. There are exceptions to this rule, in that certain firms see an advantage in playing up their environmental credentials. For example, the Body Shop has exploited a 'fair trade' policy and institutions such as the Co-operative Bank have played on their 'ethical investment' policy in order to attract investors. There has also been a rise in the number of mutual societies, LETS schemes and other co-operative groups being set up, literally forming part of the DIY culture (with a parallel process of demutualisation and amalgamation of businesses taking place). Consumers organise and can become shareholders of the organisation itself, insisting on particular operational criteria (see, for example, Leadbetter and Christie, 1999). An example that connects both examples in this chapter is the Out of This World network of organic/ethical food stores (Out of This World, 1997) as they provide ethical consumption opportunities and challenge food producers to meet organic and ethical criteria, demanding a different rural ideal. These stores operate along much the same lines as the older Co-operative Workers Society (see Birchall, 1994): the members pay an initial joining fee and have a governance system that retains an equal voting system regardless of the number of shares they hold.

Such moves imply wider provisions within a citizenship framework, including the exercise of social practices and rights and responsibilities towards the (rural) environment. Once such notional responsibilities are established, or at least aired, it is supposed that the active consumer-citizen will seek to ensure that companies/individuals adhere to their responsibilities by expressing opprobrium through his or her agency as a consumer, with its associated economic power. There are some obvious drawbacks to this view of consumer power, including the need for the consumer to have sufficient information as well as the ability to engage using the markets (read money) (see Parker, 1999c). The mechanism is imperfect; it can exclude low-income groups who may suffer the most from market exploitation or the effects of political decisions.

There are various publications aimed towards ethical consumerism and corporate citizenship that provide information for the consumer-citizen. For example, the magazines *Ethical Consumer* and *Corporate Watch* provide details of goods (and where to buy them) that meet certain ethical criteria, and report on corporate social and environmental records. They also act as intermediaries for protest groups to advertise planned actions and provide a means through which communities of interest can inform and organise themselves. There has also been a phenomenal rise in the number of smaller action groups with political/environmental agendas. The *White Book*, for example, produced by the group Brighton Justice? lists over 180 largely local environmental protest groups in 1995 (Justice?, 1995) and has been used (though perhaps now superseded by the Internet) as a means of network-building and information exchange. Such information exchange is evidenced by recent organisation of protest actions, as at Seattle in 1999 and the 2000 anti-globalisation Mayday actions staged across the world (Pickerill, 2000; Mobbs, 2000; Klein, 2000).

During the 1990s there were a number of high-profile protests over various environmental issues, some of which centred on land use, particularly in rural areas. At the forefront of these were two of the most active environmental groups in the UK (and globally): Greenpeace and Friends of the Earth (FoE). These organisations have been engaged in various campaigns involving a range of strategies concerning environmental issues. Tony Juniper, the campaigns director of FoE (quoted in Parker, 1999c), emphasises that special interest groups look to use whatever tactics can be brought to bear when engaging on a campaign. Kousis (1999: 227) outlines a range of tactics that are used by groups elsewhere in Europe. She specifically describes examples where 'communities publicized their demands through the press, filed procedural complaints to authorities, organized social events, handed in signatures, formulated public letters, took their case to court, held demonstrations and public assemblies, and conducted road blockades'. Such examples show numerous forms of 'engaged' citizenship being expressed beyond the bounds of 'due process'. Below, the Newbury Bypass protest example, one of the most celebrated UK environmental protests of the 1990s, is viewed from the consumer-citizen and rural perspective. The Newbury protest incorporated numerous tactics including an exemplification of consumer-citizen approaches to politics. Related points from the previous examples in this and prior chapters are also consolidated below.

Environment versus economic development again: urban meets rural at Newbury?

The example relates, as does the Wye Valley example in Chapter 5, to a central source of conflict in planning terms: the question of environmental protection and sustainability versus pressures for economic development. This example considers the Newbury Bypass protests which took place during 1995 and 1996 and the associated tactics and reactions surrounding the issue of a planned bypass road that were focused on by environmental groups (see also Merrick, 1996; McKay, 1998; Klein, 2000). The example is used to show how such an issue and rural/environmental backdrop were contested using market-based tactics that were removed from the actual location of the planned road and also targeted at third parties.

Newbury is a town of around 50,000 people situated in Berkshire, England. The new bypass road had been subject to a public inquiry several years prior to the protests discussed here. Newbury, in common with many other towns across the UK, had problems with traffic flows, and new road-building had been heralded as the best solution. The UK government had been carrying out a major series of new road projects during the 1980s and 1990s and the planned Newbury Bypass was one such development. The 1989 UK White Paper *Roads for Prosperity*, which was subsequently enacted, doubled the road-building programme in the UK, adding an extra £6 billion of investment into the roads network (Department of Transport, 1989). The assumptions made by the Department of Transport in order to arrive at this assessment were almost immediately challenged, with the Twyford Down road project in 1992 being the first notable clash (see Merrick, 1996; Anderson, 1997). However, it was Newbury in 1995–96 that arguably represented the apotheosis of environmental roads protest in the UK thus far. The demonstrations and protests concerning Newbury's proposed new bypass were high-profile events with widespread and prolonged coverage by the media. The 'battles for the trees' (Merrick, 1996) at Newbury offered a media spectacle that provided a focus for the debate concerning road-building and transport issues more generally, bringing environmental issues

Plate 7.1 Newbury Bypass rally, 1995. Banner depicting a politician, fed by
tobacco tax and protected by public order legislation, 'laying' new road.
(*Photo*: Gavin Parker)

to the top of the political agenda in the UK. This example, centring on a
rural location, illustrates how protest groups, as an indirect way of influencing the UK government road-building policy, targeted the companies
involved in road-building.

The protest concerned the wider principle of building new roads as
opposed to exploring and valuing alternatives more exhaustively. One of the
issues raised was that the road plan had not undergone an environmental
impact assessment (EIA) (see Glasson *et al.*, 1999; Barrow, 1997), primarily
as the legal requirement for such an assessment was not in force at the time
(EIAs were required under an EU directive of 1988; see Hughes, 1996).
This procedure was well established by the time the road programme reached
the Newbury project but was omitted, even though the finalised road route
was to pass across two SSSIs (Sites of Special Scientific Interest, the Rivers
Kennet and Lambourn). Other features defined by the environmental lobby
as being ecologically important were highlighted and aided the protest.

These included the Desmoulin Whorl Snail, a protected species. These were discovered after work on the route had begun. The tiny snails became a cause célèbre for the protesters and provide a nice example of the agency of a non-human entity in terms of network analysis and 'heterogeneous engineering': the bringing together of disparate and sometimes distant artefacts to pursue particular agendas (see Law, 1994; Murdoch, 1998). Their discovery and deployment halted work on the road for some time (Highways Agency, 1996).

The protests and actions did not all take place in and around the town itself and were diverse in nature. They included non-violent direct action (NVDA), media stunts, legal challenges, public marches and, centrally here, the use of a variety of consumer-citizenship actions. As the Newbury protest evolved, it became apparent that the sitting Conservative government would not heed the protest at stage one of the project – the clearance of the route – and reconsider its position over the new road. The work proceeded, albeit slowly and at great expense. This stage became the focus for direct action, with activists using a range of NVDA tactics including tree camps to hinder progress (see Merrick, 1996 for details). This 'battle' involved six hundred security guards and a heavy police presence, eventually costing the public around £12 million (Highways Agency, 1997). After the clearance of the route had begun, FoE put new protest strategies into operation. The direct action approach had worked to the extent that progress had been slowed down. The costs of the route clearance had been increased and a great deal of publicity had been generated, but further measures could be invoked to block the progress of the road. And in the longer term, other such civil engineering projects were halted. These included the development of a strategy that utilised the market and enabled a larger number of people to be enrolled into playing a part in the protest.

Instead, FoE devised a different strategy, one that was directly aimed at the companies involved in the tendering process for the Newbury contract as a means of influencing government as well as the individual companies. The road-builders that were placed on the tender list for the road-building contract for the Newbury bypass were a logical target for protest. However, as road-building/civil engineering firms do not produce mass consumption 'products' as such, it is difficult for the wider public directly to declare political choice as consumers. Supporters of the protest were urged, by Friends of the Earth, to engage the companies on their own ground: to buy shares in the companies involved and write to those companies, to attend the general shareholders' meetings and to make representations to the companies through 'conventional' channels.

This combination of tactics forced one company, Tarmac, to pull out from the tendering process (see Brown, 1996a, b). Sir John Banham, the chairman of Tarmac at the time, said that Tarmac was 'taking the rap for an incompetent government' when it was targeted by environmental groups over the Newbury Bypass (*Money Programme*, 1996; Brown, 1996a). He meant that the company felt that, as a commercial enterprise, it was not equipped, nor should it be, to deal with such protest actions and therefore was uneasy about having to do so. Individual corporations had been unused to making decisions that were not based solely on economic criteria. Essentially they were being exposed to unfamiliar opprobrium stemming from the 'privatisation of responsibility' (see Bauman, 1992, 1999) and felt that such actions should be faced by those deciding on UK transport policy. Of course, in the case of the Newbury protest it is the government itself that is the alleged wrongdoer and the road-building company simply the government's agent. This makes the Newbury example interesting, largely because the road-building companies were caught between the state and the environmental protesters, two groups operating with contrasting value systems.

The Newbury Bypass contract, to the value of £73.7 million, was awarded to another company, the Costain Group. Subsequently Costain was forced to hold an extraordinary general meeting of shareholders as a result of shareholder protest in September 1996 as part of the FoE strategy of using market/'insider' tactics. The meeting was attended by between 300 and 350 protesters, although according to Friends of the Earth 506 of their supporters had actually become shareholders of the company since the beginning of the campaign earlier in the year and were therefore eligible to attend the meeting (FoE, 1997). The action caused the company chairman, Lord Benson, publicly to promise to rethink Costain's involvement with the road. This event and the statement generated yet more publicity and media attention for the anti-roads lobby (FoE, 1997). It did not eventually result in Costain pulling out of the contract, but it has, like Tarmac, set up various measures to placate the environmental lobby, including an environmental forum (Costain Group, 1997a). The Newbury protest also caused the company some financial discomfort. In 1996–97, Costain shares were suspended from the stock market and doubt was cast over Costain's ability to complete the contract (FoE, 1997). Such methods of using market 'rules' are an example of how groups can use the discourse of the market. FoE had attempted to persuade Costain and the other tenderers, using these means, that the Newbury project would be untenable in economic, political and public relations terms, to marked effect. The organisers of the Newbury protest urged political action through the market in a way not witnessed before in the UK.

The road was eventually completed in 1999. Although the protests had failed to stop the Newbury Bypass from being constructed, these and similar protests (such as the earlier Twyford Down actions in 1992; see Monbiot, 1998) left a legacy. The Labour government, with John Prescott as minister with joint responsibility for the environment, transport and the regions, shelved plans for many new roads, including the proposed Salisbury bypass. In 1997 this bypass project was set to become a site of intense confrontation, with even more organised and extensive protests planned than at Newbury (see Turnbull, 1996). The withdrawal of the roads programme was a political shift that the various roads protests can claim to have played a significant part in bringing about. Similarly, the protests have ensured that environmental issues and mitigating measures are carefully worked into such schemes (see Highways Agency, 1996; Sleep, 1997; Costain Group, 1997b).

However, given recent announcements that funding for the UK roads programme is to be reinstated, the proposed Salisbury Bypass may now go ahead and it appears that a confrontation like that at Newbury would occur. How such protest is planned and executed is likely to be illustrative of the 'unprecedentedly professional' campaigns that are being waged over similar developments in the UK and Europe (see Minton, 1999; Kousis, 1999; Klein, 2000). These seem likely to include the use of market-based consumer-citizen actions in defence of the countryside, or aspects of the 'imagined' countryside against other priorities.

Conclusion: mobile politics, consumerism and the rural

Representative democracy on the present UK model has been shown to be only as adequate as its representatives are prepared to be reflexive and incorporate a plurality of viewpoints or positions. The development of politicised consumer tactics reflects attempts to short-circuit political rights where government is viewed as being slow, unresponsive or concerned only with the needs of the powerful or of the majority. In that sense, people are using consumer rights to voice citizen interests. In terms of the countryside, people are more aware than ever of their rights and have a developed sense of how others in the countryside should also behave. They act as consumers as much as citizens in expressing how the countryside should be managed and used. In this latter sense, people are making use of citizen rights to protect their interests as consumers.

The Countryside Alliance has adopted the tactics and rhetoric of wronged consumers and marginalised citizens in its attempts to defend hunting rights and highlight other rural issues. Consumer-citizen tactics provide an opportunity for the public to influence decisions actively in at least three ways: over state action *per se* (e.g. the Countryside March and Rally); when targeting agents to influence the power enabling the cause of the conflict (e.g. Newbury); and over the actions of an alleged or possible antagonist directly (e.g. Nestlé).

Both examples in this chapter provide illustrations of Albrow's (1996) notion of 'performing the state'. Particular semiotic and iconographic representations of the rural and the environment for or against particular groups were used to influence government on matters of policy – in particular, and in the light of the foregoing details, the way that the countryside is constructed and contested in the era of lifestyle-politics and how single-issue politics in the context of consumption and fluid space and mobile citizenship are orchestrated and presented. Klein (2000), when examining the tactics of Reclaim the Streets, talks in this respect of 'large-scale coincidences' and actions 'arriving'.

Chapter 5 introduced the notion of the discourse of heritage and outlined how it is being drawn upon, even to the extent of its being appropriated into the discourse of stewardship used by established institutions such as the Country Landowners Association and the National Farmers Union. This competing, and perhaps complementary, discourse of heritage brings with it a rather different ethic – not necessarily of 'leave it to us; we know best', but rather a preservationist stance that seeks to displace the stewards with a managerialism (and a territoriality) that selectively seeks to fix tangible aspects of 'rurality': for example, the built heritage or landscape character over practice and activity that have been subject to (perhaps laggardly) change.

It has been argued here that the market – the discourse and practices of the market, within capitalist consumer democracy – has been playing an enhanced part in the politics of protest in the 1990s. There is little doubt that consumer-based protest does take place and that it may provide a useful platform from which to criticise alleged wrongdoers or to highlight alleged bad practices. The area of politicised consumption certainly provides useful subject matter for further research. It is unlikely, however, that the types of market-based consumer-citizenship outlined here can achieve systemic change on their own. Rather, consumer-citizenship is becoming a more commonplace tactic which citizens incorporate as a form of political right-cum-responsibility, added to a range of other processual and formal political actions. This results in interest groups, comprising active consumer-citizens, using the marketplace or related institutional routes, such as boycotts and shareholders' meetings, to register protest (as shown again through the fuel blockades of September 2000 in the UK and across Europe).

There are also instances where such tactics are shown to be ineffective and inadequate replacements for effective regulation by government. The pan-European action over fuel prices took place only a matter of weeks before a 'Dump the Pump' campaign promoted by pro-road user groups. Consumers were exhorted to refrain from filling up with petrol, but they had little alternative but to do so (Dump the Pump, 2000; *Guardian*, 2000a). Another instance relating to oil companies and fuel was the boycotting tactic used against Shell as part of the Brent Spar platform dumping campaign. This was rather different and more effective because it was focused on one oil company, thus giving consumers the room to buy petrol from alternative sources (see Parker 1999c; Macnaghten and Urry, 1998).

There are some fundamental drawbacks with consumer action *qua* political action in terms of effectiveness, democratisation and political power that warrant mention here. First, the market determines the form of protest and tends to control the information required to make informed choice concerning the behaviour of market players, and information regarding the behaviour of companies will reach the public domain infrequently. Therefore 'protest' concerning the behaviour of market players is tightly controlled. In many cases such protest is not a sufficient remedy because the damage being indicated is likely to be but one instance of potentially more wide-spread problems. Second, the media as messengers of protest groups and as investigative agents in their own right are unlikely to provide consistent and reliable sources of information, tending rather to light upon issues in an *ad hoc* manner. They are themselves unlikely to be able to determine the relative merits of particular (constructed) arguments. Third, as the environmental lobby argues, this incremental process does not protect unique or fragile environmental goods that, once lost, cannot be replaced or be used to remedy other irreversible wrongs. Instead, *ad hoc* whistle-blowing or intermediary action on the part of other actors, for example supermarkets or consumer magazines, will continue as useful but inefficient market sentinels. At the other pole, groups with resources can use market-based tactics to raise the profile of particular issues by disrupting economic activity.

Such strategies or modes of expression are closed or inaccessible to some groups and can add to or compound social and political exclusion. For example, people without email or other high-technology innovations are disadvantaged. Although not central here, such points lend weight to the argument that citizenship in all forms is contingent, as with factors of exclusion/inclusion and poverty/wealth. Notwithstanding such criticisms, consumer-citizenship appears to be an important approach or tactic in contemporary politics that can be used in various contexts (see, for example, Welford, 1997). It may be argued that multiple modes and opportunities for engagement enable a widening of participation and politicisation. The philosophy and rhetoric of the 'active' citizen as encouraged by both the Conservative and 'New' Labour in the UK might view the use of the market as a legitimate political instrument (see Chapter 3). The consumer-citizen tactics described largely represent legal and 'non-deviant' methods of registering protest through active and engaged citizenship. They do, however, strike at the heart of bureaucratic process and formal modes of public participation, hence the litany of public order legislation passed, though not always enforced, over the past fifteen years.

The development of consumer-citizen tactics means that a fluid mix of the state and of private interests and the media is regulating the interests of the citizen and the consumer. Ever-tightening controls over 'protest' raise important issues concerning available means of transformative action (see Sibley, 1995). The development of consumer tactics in the way described, the limits placed on dissent and the treatment of minorities through state policy bring into focus questions about the future of representative democracy in the UK.

The paradox is that few want change, not even the stewards, but they have always responded to (bounded) economic conditions. Traditional interests lose twice over, both in terms of production providing political clout and as a result of widening understandings that farmers and landowners have not always prioritised heritage or environmental issues. This situation brings into sharp relief the role of the state in determining legitimate rights and responsibilities, and how such processes of legitimation are affected by particular constructions of reality. Pressure is placed on government and others, by interest groups and the media, to incorporate change. The development of strategies on the part of the public to have further influence over decision-making within the public and private spheres and beyond the representative democratic process is therefore, perhaps, unsurprising. Further, it is also a reflection of the dominance of markets and commodities and the importance of knowledge that citizenship, for many in advanced industrial societies, *is* consumption, both of goods and of information, rather than direct political engagement.

Government needs to be aware of the diverse and fragmentary ways in which 'citizenship' is being expressed and to allow such diversity to be reflected in terms of rights distributions and mechanisms for garnering opinion. The state will be required to assess rights and responsibilities not only in relation to private property, animals or the environment, but also in terms of diverse and perhaps localised cultural and political rights. By association, the way that consumers may play an increasing role in terms of governance or governmentality – the 'conduct of conduct' (see Simons, 1995; Foucault, 1977) – implies the need for a renegotiation and redefinition of what constitutes 'active' citizenship, and perhaps incorporation by states to respond to it creatively. The forms described here illustrate that the development of a more inclusive governance will require a radical restructuring of the way in which decisions are reached in both UK politics and private business in the future. The media bear a large and potentially inappropriate responsibility in terms of presenting consumers *qua* citizens with information and potentials for empowerment. Conceptually, insider forums and the like still offer citizens and protest group(s) bounded opportunity, or, in Sack's (1993) terms, enable them to operate within the structures imposed upon us and, in Foucault's terms, to 'step over the threshold of the discourse' and be enrolled by the oppositional interests. Such strategies can effectively validate the decision-making process that is already in train (see Foucault, 1988; Murdoch and Abram, 1998; Forester, 1999).

Chapter 8 concludes with a discussion of the post-rural and the increasingly commodified countryside, including how they are likely to be shaped in the future by such actions and activities on the part of disparate, possibly incongruous, groups and individuals. This closing chapter also speculates on issues that may become more important in the coming years in relation to land reform and regulation, the formulation of new communities, and attempts to shore up and protect rural localities from global economic pressures. It also examines how, concomitantly, other cultural and political changes may (continue to) transform both the countryside and the way that citizenship is enacted and received, regardless of the best efforts of national or local government or established power elites. More darkly, it looks at how powerful groups can orchestrate destabilising actions using a range of 'cloaking' instruments drawn from a diversity of sources.

8 Citizenships, contingency and the countryside

Multiple, contingent and inclusionary citizenships?

While tying the text together by reviewing the salient points raised or implied in the preceding chapters, this conclusion aims to look at the relevance of citizenship in the 'post-rural' and the 'post-national' contexts detailed previously. In Chapter 2, citizenship as a concept is examined in some detail, and the recent use of citizenship by UK governments is discussed in Chapter 3. The historical commentary provided links the first parts of the text, giving a contextual base from which to develop and explore contemporary examples in the later chapters. Here, the increasingly commodified rural and the complex, interconnected network of relations that impact on the rural are discussed in relation to the theory and examples introduced. The methods or current ways of encouraging and engaging citizens are discussed in relation to a widened view of the constitution of citizenship.

A more expansive view of citizenship than is traditionally considered or applied in rural studies, human geography, land use planning or related disciplines has been adopted in this text. This is partly a reflection of a changing cultural, ecological and political environment. In cultural and environmental studies a wider view has started to be taken of citizenship (see, for example, Day, 1998; Stevenson, 2001), but the implications for such a widened definition (one that includes direct action, for example) are not fully elaborated. For example, 'participation' and 'action' are implicit in a range of rural and planning texts as constituents of citizenship, but not necessarily expressed as such. This approach towards citizenship means, of course, that a wide range of actions and practices may be seen as part of 'citizenship' and that some of these aspects are in opposition to state policy, or the positions of elected authority, and may be directed at different scales of governance. Hence part of the overall thesis relates to the structuring effects of agency in and on the countryside. One of the stabilising factors is rights and responsibility distributions, with concomitant destabilising counter-claims and practices.

Ironically, the partiality of this work is made more acute by the widened boundaries of the conceptual frame and in terms of its consideration of land

and the rural. It has been beyond the capacity of this text to examine the range of other issues, activities, policies and influences that are important adjuncts to the debates, concepts and points raised. For example, the issues of exclusion/inclusion and public participation/non-participation in policy debates extend into many contexts. Many gaps and strands provide clear areas for further research into a range of citizen actions and the defence of rights and claims. In particular, attention might be turned more fully towards practices, customs and nascent rights in terms of local citizenship. Additionally, questions can be posed regarding how citizenship, in the traditional sense, is relevant in the postmodern context. How can a broader understanding, acceptance and reflexive application of citizenship as process, as activity and as contingent identity be incorporated in terms of governance, law and policy? What is the legitimate role of the media in politics? How can national governments better regulate and incorporate wider examples of citizen action? How can government and other actors move to empower a wider constituency of citizens so that they can engage critically with an increasingly complex and multiscaled political economy?

While the book does begin to address such points, there are also further questions raised for research and theorisation in this respect:

- What are the consequences of postmodern/post-national citizenship(s) for traditional tiers and modes of governance?
- What are the objectives of citizens as activists?
- Can formal democratic systems be rescued?
- What are legitimate modes of expression? And for whom?
- How can nation-states cope with or assimilate such citizenships?
- Are 'core rights' now more about 'core morality' and therefore is the cultural overthrowing the legal?
- How can the role of planners be promoted to facilitate better citizen information, action and participation in terms of land and the environment?
- What is the 'best' use of land in the social sense (including considerations for health, education, as well as leisure and productive uses)?

Numerous changes, cultural, economic and technological, have given rise to new or newly configured practices, particularly in the past fifteen years. Economic requirements and relations at different scales have been changing, and attempts to alter power relations have accompanied such economic developments. Practices and uses of the countryside have contributed to changes in the economic and the political, with historical relations and readings of history and environmental and cultural features implicated in brokering change. Reorganisation is, however, uneven, with inconsistent outcomes in a period of transition that appears to have no endpoint. Instead, there are multiple potential trajectories and contingent stabilisations that are under increasing scrutiny and challenge from more and more interests.

The examples used in the later chapters have been resultantly diverse, yet with certain cross-cutting themes. These themes appear for three main reasons: first, as a result of the research that has informed the case studies and that centres on environmental planning and the politics of land use that underpins the book (and as such, a generative constraint); second, owing to the insight that conflict and contestation provides for analysis; and third, as a result of particular predilections or discursive tactics of postmodern political action during the late 1990s. It is to be reiterated, however, that the examples used are only the tip of the iceberg. Other examples of citizenship expression are legion and diverse, from the high-profile political to the normalised, the everyday or mundane. This, again, is a consequence of shifting the scope of citizenship definition. The old lady in Cumbria who opens her house weekly as a doctor's surgery and village post office is but one example of the latter 'good neighbour, good citizenship' rural action (see Moseley and Parker, 1998; Moseley, 2000). These forms of action are closely aligned with notions of self-help and charity that have underpinned recent Conservative and Labour conceptions of citizenship. Essentially citizenship, in this construction, implies self-reliance and responsibility being shouldered by individuals and 'communities'. This model also helps governments to evade and pass on responsibilities to groups and other institutions. This may be legitimate or justifiable in some contexts, but is it open and transparent? Is this approach towards politics an example of power without responsibility? If so, how can responsibility be matched with rights, entitlements or powers across the range of stakeholders and protagonists, including governments and corporate interests? Conversely, should the state instead be rethinking its approach towards, and the direction of, an increasingly postmodern politics?

The past twenty years of governmental citizenship rhetoric is partly responsible for a polarisation between the previously defined disillusioned non-participating majority and alternative cultures of direct and market-mediated action. Nevertheless, it can be argued that citizens are increasingly involved in politicised performance in at least two ways: first, in terms of demanding performance from authority, and second, in terms of themselves performing in different ways. A third or cumulative form of performance is identified and cited throughout the text – that is, the way in which citizens in these two senses actually 'perform the state' (Albrow, 1996). This is where governments follow, unevenly and sometimes unaccountably, the direction of public opinion. This largely mediated and manipulated, or conversely sectarian, opinion uses the media to manipulate governments and other powerful institutions. These forms do, however, require further research to explore people's motivations, tactics, networks and the outcomes of their actions – particularly over longer time frames.

If citizenship is a shifting identity (and status) that comprises a wide range of practices, and further, if citizenship concerns are being expressed beyond the national, then how should transnational organisations respond and involve global citizens in decision-making and the brokerage of global rights and responsibilities? This question is extremely complex to mediate, and empowerment does in some sense mean bearing a form of collective responsibility, or being 'community-regarding' in Sagoff's (1988) terms. In a rural application this, for example, relates to impacts on agriculture due to global commodity prices or the impacts of tourists on culture and landscapes. Different economic sectors and distant places are increasingly interdependent – as shown in the September 2000 fuel crisis that impacted across Europe (and also with the foot and mouth outbreak in the UK in 2001). It is interesting to note how the fuel crisis was begun by French farmers and haulage firms and was instigated in the UK by a group of Welsh farmers unhappy about the state of the rural (agricultural) economy, with the high price of fuel (apparently) being one factor in their plight. There have also been suggestions that the UK action may have been encouraged by other indirectly related interests keen to destabilise a Labour government likely to ban hunting and in the process of introducing a 'right to roam' (*Observer*, 2000). Such events and tactics also illustrate the increasingly interconnected nature of postmodern politics. More specifically, the politics of consumption is illustrative of the crossover between the economic and the political spheres.

Projects and practices of citizenship

In the first part of the book, three overlapping 'takes' on citizenship were set out, providing an historical, political and theoretical backcloth for the later chapters and for future research – in some senses providing a foundation for future work looking at rights, practice and citizenship in the countryside. A deeper look at how the countryside produces (after Baudrillard) 'signs' of political action – signs of citizenship – is also worthy of further exploration. A widened definition of citizenship as conceptualised here is, however, fraught with danger. It becomes so wide and diffuse that it may begin to lose clarity and explanatory power. In short, it can become little more than an excuse or rationale for explaining everything neatly as part of 'citizenship'. The word is retained here, but it seems that an alternative or prefix, such as post-citizenship, is necessary in order properly to acknowledge political and cultural change in the global age. Certainly a necessary re-evaluation opens up debate about how and why governments attempt to define and regulate citizens in the ways they do, particularly in an era of globalisation and, in the UK context, where government has been becoming more and more restrictive in terms of tightening and defining more closely what is and is not acceptable behaviour: defining what is good citizenship.

The deeper history of citizenship, nation-building and the focus of land and rights illustrates how the state has coerced the population into accepting a social contract that was first and foremost an economic contract. Arrangements concerning land use and appropriate activity, and the apportionment of rights and responsibilities when first constructed, have undoubtedly had a lasting impact, with many rights and responsibility distributions subjected to little significant change over time. The contract was legitimised as being 'rational and beneficial' and providing economic efficiencies to the wider (national) community. The feudal system, which predates the modern state, was inextricably political as well as economic. Change was portrayed as necessary for the best use of the rural as productive resource. Such an implied justification has persisted until only recently, and certainly until well after the Second World War. Only secondarily and more recently (see Chapter 2) has the contract been manifesting itself into a social arrangement to rival the economic. Gains in social rights in the post-war period and gradual globalisation of economies provide a juxtaposition of the political and economic forces that are now shaping and have previously predicated aspects of citizenship.

Expressions of citizenship lie in the cultural, historical, economic and legal. Citizens act according to particular interpretations of those codes and structures. Such interpretations also involve the brokerage of change. As discussed, even the historical is implicated where attempts to undermine extant and dominant interpretations are subjected to challenge from competing communities of interest, sometimes using alternative histories to tell stories about the future. Short (1991: 5) talks of how 'myths destroy time'; now there are examples where time is used to create myths and to uncover a logic for past relations. Such myths are used to signify and inform imagination and contemporary practice. Such political tactics may not be 'good' but are most probably healthy where old or dominant conceptions are moribund or suffer from complacency.

In the UK, Tony Blair moved towards a programme using the 'social efficiency' model in order to justify policy change at the national level. The progression of welfarist policy and protection of the countryside as productively beneficial space has been subject to challenge and internal crisis. The role of rural land and the future of social and cultural arrangements in rural areas have been moving closer to urban normalisations and are being appropriated increasingly by the wider population for consumption purposes. The rights being claimed and enforced are shifting, with strenuous attempts being made by landowners and farmers to protect and reinvent the economic contract – for example, through attempts to commodify access for recreation, and receive payment for 'good' stewardship (read citizenship), particularly in terms of environmental citizenship. Maintaining the 'national' interest, as with national citizenship, appears to require economic regulation and a form of protection for agriculture and other forms of rural economic activity. Such support would, however, further revalorise the countryside into museum space, or in some sense encourage hyper-real space. Consumption, it can be argued, may bring its own obligations. The consumer seeks optimal resource use and this search leads to conflict over such optimisation between sociations.

UK governments have attempted to pursue particularist citizenship projects, such that those projects play an important part in wider areas of governance and policy and also provide a good indication of the underlying philosophy of government. Administrations over the past twenty years (but, as is noted in Chapter 2, over a longer time span, since the formation of modern nation-states) have raised the stakes in terms of spatial regulation of misdemeanours such as trespass and in terms of surveillance through a range of policies and legislation. These government actions represent the disciplinary 'threat' whereby the state draws a line of demarcation to contain political action – with trespass being part of that rhetorical threat (see also Shoard, 1999; Parker and Ravenscroft, 1999). High-profile examples of government attempts to do so include the Criminal Justice and Public Order Act 1994, the Investigatory Powers Act 2000 and the Terrorism Act 2000 (Parker, 1999a; Morton, 1994; Home Office, 2000). Terrorism is now legally defined such that action against property may be classed as terrorist activity under the legislation. Action that is considered to be ideologically or politically driven can similarly be defined as 'terrorist'. It is noted that such powers are deployed and enforced inconsistently, and 'rights' appear and disappear continually in such circumstances, a fact that highlights calls for a bill of rights so that particular 'core' rights (see Dahrendorf, 1994; Charter88, 2000) are in place to curb government power. These repetitive flourishes of disciplinary legislation may also be read as the actions of national governments that are increasingly threatened by pressures from above and below, outside and inside their domain. Accordingly, the discourse of citizenship adopted by governments in the UK is both elegant and pliable. Put more unkindly, it is political spin that is inconsistent in form and substance. One challenge for national (and regional) politicians is to engage reflexively with political and economic changes taking place, and being demanded, in the global and post-national age – again suggesting reappraisal of political stances taken towards domestic industrial activity, and, with the Human Rights Act 1998 in mind, how 'uniformity' of treatment is demanded by a plural 'society' (Parker, 2001).

The notion of the 'gardening state' and the 'gamekeeper state' (Urry, 2000; after Bauman, 1987) is a good way of conceptualising what may be occurring in terms of citizenship, land and the countryside. I acknowledge that this general statement should be treated with caution, but would argue that disciplinary approaches are taken by the state, particularly perhaps under conditions of postmodernity and under the pressures of globalising forces. Land is a multiple resource and there are multiple potential benefit streams. This implies a range of potential conflicts over rights and responsibilities and activities in rural space, but there may be ways of minimising such conflict and maximising the potential of the land as site of consumption and production, preservation and development. Unfortunately, land use and ownership are tied to cultural and economic inertias that have more or less ensured path dependency. Such factors also hinder wider and socially efficient land use.

Land is an anchor of citizenship as identity, status and activity, and it also becomes the platform and focus for a range of interconnected conflicts, each affecting citizenship in some sense. Trespass has been used as a key example in this regard, where citizenship rights and responsibilities are clearly defined but not necessarily enforced. The approach taken here has included citizenship as practice, a stance that appears to parallel that of government exhortation in terms of active citizenship, yet this call from successive administrations has been strictly bounded and has indirectly reinforced exclusion, for example through law and order policy. It is also to be debated whether much effective change in systems of democracy, despite devolvement of some powers to Scottish and Welsh assemblies, has been achieved or is being countenanced by government. Thus a broader conceptualisation lends weight to calls for the broadening and deepening of the engagement of citizens in policy and decision-making processes in the UK and elsewhere.

It is also the case, reinvoking Van Gunsteren (1994), that each (government) action can be read as affecting citizenship. Such actions are inconsistent and represent trades, bargains and manipulations that are attempts to steer society and placate powerful interests. For example, farmers can be told to be good stewards and also be paid to allow recreational access, which presumably requires a trade-off between economic and social efficiency. By the same token, other groups can be persuaded to be good citizens *à la* P3 example by fixing stiles and gates and clearing paths regardless of legal responsibility and without recompense. Such stances reflect the different priorities and imaginations of the rural that need to be reconciled.

Simmie (1974), in a seminal 1970s text, saw that planners were an important group attempting to mediate power and conflicts on behalf of a range of groups and interests (and seek best use of land). Some twenty-five years on, it might be argued that only a small proportion of citizens engage politically – albeit in novel and diverse ways – but that many more through their practices are involved indirectly in (micro)politics. On a more philosophical level, all actions are in some sense political (Van Gunsteren, 1994, 1998). Those who do engage tend to be reactive and reflect issue-based communities of interest campaigning for common causes. Less political, but still intimated in the everyday politics of practice, are the volunteers of different types who practise benevolent or charitable 'good' citizenship. It is still clear that many are excluded from engagement by a lack of different forms of capital – and not only the economically excluded, as implied by writers such as Simmie, but also the culturally, socially and legally excluded.

Different modes of engagement, some already practised and outlined beforehand, should be recognised as legitimate and important expressions of citizenship. This is particularly so given the impacts of globalisation, the communications age and the dominance of a culture of consumption. In short, this relates to environmental and rural agendas in that there are different ways of valuing the countryside and its features, practices and difference in terms of local culture and globalised networks of relations. There does need to be further work done to better understand and more effectively encourage citizenship action, as well as engendering a public debate about UK systems of governance and engagement/empowerment at local and global scales. Such a debate should include the way that children (and adults) are informed and educated about politics and the potential role and power of the individual in (global) society. In this sense there is a need to recognise more explicitly the connection between citizenship and the study of social/human capital. It might also involve a re-engagement with the land and the wider rural environment in order to help close the gaps between urban and rural knowledges or competencies. Many people are alienated from the land and have a somewhat ersatz view of the countryside; not necessarily a correct one, but a pervasive view often based on little personal experience. It is a gaze readily packaged, simplified and mediated for consumption. The effect in terms of citizenship is to produce a range of actions and views, some of which inform and consolidate 'imagined countrysides'. One key message should be that systems, policies and power relations are contingent – as indeed, therefore, are the status and configuration of 'citizenship' itself. One of the points implied is the effect of actions and imaginations in terms of the structuring effects of such agency.

The ways in which economic restructuring is reflected in present government policy and the influence of supra-national forces, such as the European Union, are central to the trajectory of the rural economy and society, and more widely to citizenship. They are also seen in the way in which landowning and farming interests have reacted to economic change. Many rural commentators have identified the main single economic factor behind the restructuring of the countryside as the influence of the Common Agricultural Policy and the ways in which this policy is being redesigned (see Winter, 1996; Thirsk, 1999). The rural is changing, however, because of other factors of change, many of which governments do not appear able to control. Society and its constituent groups might easily rival claims of civic sclerosis, made on the part of the populations who mandate governments, with claims of political paralysis on the part of their elected representatives. Change is inevitable; it is how it is debated, moulded, mediated and accommodated that is in question. The coming of a post-national society might be a true time for the citizen, in his or her many guises, to emerge and be recognised as a multifaceted political agent. The emergence of such a citizen is desirable in order to democratise and pluralise governance and the mediation of cultural, social and economic change.

Consumers, commodities and the rural

The role of land, traditionally as site of production, has been witness to a shift towards consumption, and towards increasing suburbanisation and associated 'conservation' of the countryside. This alters trajectories and sets the erosion and defence or reinforcement of rights and responsibilities (renewed through practice and related actions such as protest or 'story-telling'). Land is being increasingly consumed in a variety of ways, and alternative exchange values are being developed, with different elements of land, and its use, being commodified. Commodification alters social as well as economic relations, including aspects of 'community' or elements of trust, sets of ethical concerns or moral components (as mentioned in the Countryside March example).

It appears logical that political identity is, and perhaps in some sense should be, mediated by consumption (in the wider sense of the consumption of space as well as relating to perishable or hard-goods commodities), given that many flows of postmodern society operate through consumption and its mediation via the sign economy. The rural can be examined in terms of the role of semiotics in cultural terms and therefore in political argumentation. Semiotics and the sign economy (Baudrillard, 1988) leadchange and cultural imaginations. Lifestyle shifts in terms of consumption and the politics of consumption require further consideration in terms of the impacts of lifestyle politics and the countryside. The Countryside March, the Wye Valley and the P3 examples, in particular, highlighted the defence of consumption practices in spite of opposition and legal challenge. It is suggested that communities of interest are increasingly coalescing to defend and claim consumption rights, a shift from earlier concern with the politics of work and production that have dominated British (rural) politics. A variety of discourses were deployed in order to legitimise groups and the activities or practices that they wished to pursue or, conversely, see outlawed. Rose (1990) sees processes of commodification and consumption practice as part of building and being offered identities in identikit fashion. When Rose's ideas are applied to thecountryside, what impacts and reflexivities might be implied? Media and consumption practices shape the countryside as surely as the countryside and its mediation have shaped citizenships. In a number of the examples used in this book, it was symbolic capital that was deployed, rather than economic. This was mobilised in order to maintain exclusion as much as to gain 'entry' or maintain inclusion, thus again highlighting the inclusion/exclusion issue.

Commodification is a process of marketising practice, culture and resources in new ways or for new markets. Monetisation erodes alternative means of exchange and social arrangements. While the market is becoming a powerful tool for protesters, it is being regulated at least as much by big corporate interests as by national governments. This partly explains the rise of consumer-citizen protests and pleas for good corporate citizenship (see McIntosh *et al.*, 1998). However, while governments attempt to define and curtail 'citizenship' for populations within their territories, they appear to be able to do little to regulate the actions of and define citizenship effectively for transnational companies. This is particularly so when such companies hold powerful bargaining chips that can threaten politicians with electoral defeat, as in the case of the threat of thousands of redundancies. It is possible to argue that if citizens are attempting to perform the state, then the corporations themselves should be seen as part of the state as well as elected government, or entrenched bureaucracy and other forms of authority.

It has been argued that the landscape in the most general sense is intrinsically a landscape of semiotics, being geared towards consumption. Not only is the landscape constructed, but also the notion of citizenship and of legitimate or acceptable behaviour is socially constructed and culturally defined. It is likely that the public (via different media forms) will become more (albeit partially) informed than before and will look increasingly to 'perform' the rural as consumers of produce, consumers of landscape and through (primarily leisure) activities. Consumption of these types is leading us steadily towards a politics of consumption by consumption.

Drives towards extensified leisure practice, in particular current debates over issues of hunting and walking, may represent important features of a commodified landscape and rural 'space' – a space that is at once multiple and diverse, but is also contested at the national and global levels. This is where particular issues or activities become the visible conflicts through which different classes and other societal fragments manoeuvre for rights and benefits, imagined and real, symbolic or exercised. The Countryside March example was used to introduce the ongoing struggle taking place between the powerful landed interests and the culturally, discursively powerful middle class – constructed by the Countryside Alliance as being between an informed rural class and an unskilled urban political class. Interestingly, the march and rally (as with many other conflicts and protests) have been prismatically contested through the defence and claims towards consumption/leisure practices and associated rights.

The countryside as experienced by most 'outsiders', and as imagined by many living there too, is one of signs and symbols, of 'brands' and labels indicating why the countryside should be valued – for example, on the grounds of the signification of history, heritage and amenity value. Examples of such branding lie in the oakleaves and the portcullis of the National Trust and English Heritage respectively, and in the labels devised to interpret place, including 'Shakespeare's Country' and 'Hardy Country', seen adorning the portals of particular favoured and encultured areas (see Crang, 1998; Rojek and Urry, 1997). The countryside is increasingly a cumulation of overlapping, 'branded' and otherwise designated patches (see, for example, Macnaghten and Urry, 1998). There are public designations and labels aplenty: national parks, AONBs, country parks, national nature reserves, SSSIs. There are semi-commodified 'chains' of locations run by English Heritage, Scottish Natural Heritage and the National Trust. Each 'badge' the countryside, and a particular extract of 'heritage' is put forward as the appropriate consumption experience. Visitors experience the countryside in particular ways and yet are tutored; they come to expect degrees of particularistic homogeneity in land management and practice.

An amusing yet indicative artefact of this presentationalism is a 'brown sign' on the outskirts of Cheltenham – on one of the roads that takes the

Plate 8.1 'End of scenic route' brown sign on the outskirts of Cheltenham, Gloucestershire. (*Photo*: Gavin Parker)

traveller up onto the Cotswolds ridge and, in the other direction, into the 'Regency Town' itself. It says simply 'end of scenic route', thus indicating how one might retrospectively wish to prioritise the experience of the rural that one has just had (or is supposed to have had) over that of the town. Moreover, it indicates that one needs to be told that the scenery or the show is over and the next brand is being encountered. This may be seen as a new form of enclosure: by commodification and fetishisation. It sets up prerequisite boundaries and contours for new citizenships and 'new' territories.

Similarly, there are privately run brands that sell themselves and sell the countryside – simultaneously selling images and imaginations of the countryside. There are many commodities unrelated to or tenuously associated with the countryside that use and feed the imagination of particular countrysides. Meanwhile, producers and others traditionally engaged with businesses associated with the rural are affected. Such processes affect citizenship at both the local and the national level. In terms of citizenship, actions may be increasingly taken through channels of least resistance – that is, through actions as consumers of goods, services and places. Attempts are being made explicitly to tie together the productive, the historical, the cultural and the consumptive modes of experience of the countryside.

A logical progression of the above has been the recent MAFF (now part of DEFRA) launch of the 'Eat the View' campaign (MAFF, 2000). This explicitly links food products to the aesthetic qualities of rural areas – a fusion of two forms of consumption. Partly the promotion is intended to encourage local produce to be consumed locally, retain economic value locally and cut down 'food miles'.

The attitude of many citizens (as consumer-citizens) is to pass on responsibility for the countryside to groups such as the National Trust, CPRE, and Friends of the Earth. They buy action via the expression of consumer-citizenship by funding political action by representative groups. The countryside is being valorised as part of a trend towards social, if not cultural, ownership of the land (see Bromley, 1998; Parker and Ravenscroft, 2001). It is also part of the drive that feeds and is fed by the heritage discourse. The bundling and commodification of time may be another area where further research should be carried out in terms of countryside use and political action, one aspect being how people make decisions about their actions and inaction and how people manage their time and consumption patterns in the widest sense. History and heritage are important components of the commodification of time and are used to control citizenships. This begs a question for further research: to discover how strategic people are in these forms of 'citizenship' expression.

Administrations view citizenship and indeed heritage relatively narrowly. Along with the closing of definitions in terms of appropriate and legitimate behaviour come policy vehicles such as the P3 scheme (as well as a host of other schemes and policy vehicles that could be analysed in much the same way as in Chapter 6). There is a range of alternative strategies that citizens can employ in order to further their rights-claims. The actions and rights of citizens in the UK (and both within and beyond the frame of the 'national') are diffuse, diverse and contingent. Thus the views and actions of many groups and sociations add to a plurality that is a constitutive element of postmodern citizenship. Property rights, and by association citizenship rights, are distributed and regulated in particular and contingent ways – rights are continually being redefined and shifted by legal process, by cultural change, and by the economic shifts dictated by market measurements. Such change dictates and is dictated by a whole range of cultural, social, political and, of course, economic factors.

Agency, participation and the rural

Formal participation in the policy process is constructed as 'good' citizenship. To an extent, however, citizen participation is structured by powerful interests as part of wider processes of political, economic and social change. Discerning legitimate and 'good' citizenship is far more complex than governments appear willing to concede. In associated form, individuals are used by other group interests to promote particular claims, as for example lorry drivers being told to blockade oil refineries by their haulage company boss (and possibly other interests), or landowners paying estate workers to protest. In the former example it might have been more active and engaged of the driver to refuse. (Yet more significant in that particular action were the oil tanker drivers who refused to drive their vehicles, thus participating in the protest through their non-action.)

New forms of political engagement have spread, including the growth in membership and diversity of the environmental movement. Recent years have also seen the rise of direct action protests, mass consumer boycotts, and an increasing scepticism about the ability of conventional areas of expertise (e.g. scientists and government officials) to inform public policy decisions, especially in managing risks to public health and the natural environment. This situation leads one to question what exactly is wrong with existing systems of governance. Is devolution using systems that mirror that of Westminster sufficient in the post-national formulation of autonomous regions and districts? What of extending the principle of subsidiarity? New ways of thinking about these issues, and perhaps more transparently, should be encouraged. Is there a case that forms of radical, even quasi-anarchist, modes of governance need more open discussion and credence given over to some of the ideas that underpin such philosophies? While this text cannot possibly delve any deeper into these issues, it is suggested here that no change is not an option for national governments.

Citizenship and change

One consequence of the application of a wider citizenship frame is the acceptance of contingency. Citizenship may be said to be the product of change as much as it is about the punctuation of change with legal or cultural rights and responsibilities. Groups use citizenship discourse to advocate for and against change. This allows a more panoramic view of the range of spatial and temporal influences that have impacted on citizenships. At least four linked drivers, at different spatial and temporal scales, affect the operation and trajectory of citizenship. These are history and heritage, prevailing economic requirements and associated land use, practice and performance, and the exercise of differing forms of power. The examples used in this book provide an insight into those factors, signalling how little research has been conducted relating to the constructions and expressions of complex interconnected processes and discursive interactions. They also show how citizenship may be seen to fragment and multiply; this engenders internal and external (using the national as the base scale) conflict. At the level of the individual, citizenship may be said to be the experience of change and reactions or constrained reaction to change by the individual. Such experience implicates status, identity and activity; who, why and how people claim or enforce rights is part of a postmodern politics of citizenship. This latter point is also worth further exploration to examine how people's claiming or enforcement of rights is effected.

In the longer term, more flexible and localised decision-making, or at least deeper public participation, should be encouraged and enabled. While on the surface this is being promulgated by government, the issue remains, how can people be motivated and their 'capacity' built up? How can politicians and policy-makers better take into account practices and activities already performed? What are the most effective ways of ensuring healthy and informed dialogue? In terms of the rural as topic, wider interests and views should be welcomed in such dialogues.

There is little doubt that some of the impacts and processes of globalisation are impossible to reverse, nor is it desirable for them to be reversed. It is, however, possible even within a modernist, if evolving, 'super-state' such as the European Union to seek to preserve rural services, local economies and aspects of local culture. The way to view change at all scales is as a challenge for all levels and agents of governance and for those with imaginative and useful ideas outside of governing institutions to voice their views. This is likely to ensure that change can be brokered advantageously, or, at the least, regulated to minimise outcomes that are considered most unwelcome. It does not mean that relations between groups, governments, producers and others will alter. The role of the national state should alter to become more facilitatory: providing education and acting as a conduit for information exchange between different scales.

There are complicated issues, in terms of legitimating rights and requiring responsibilities, that are problematic in applying a citizenship frame to the needs of a more dispersed, mobile and affluent population (coupled with wider processes of globalisation and global change). They raise acute challenges for different social and economic groups and can generate high levels of anxiety, insecurity and conflict. One of the arguments made in this text is that citizenship should be widened to catch up with how people express their opinions and how culture leads (and should be seen in that light: more transparently) the political and the legal fields. Participation in formal politics in the UK is at a very low ebb, but politics in the widened sense certainly is not, with all types of protests and actions being staged. Such actions are targeted around the world at different scales: for example, companies, governments, and in relation to particular practices that morally offend groups and cultures – of itself a consequence and process of globalisation.

One effect is the apparent dilution of concentrated effort or concern about particular issues. As mentioned above, one important response has been to turn to 'memberships' for 'sociate politics' and the market as ethical consumer, or to the symbolic gesture of the media stunt or other spectacle. Given that a majority appear not to 'engage' in normative forms of political activity, different ways of enabling engagement should be considered and encouraged. Effective active citizenship requires information and education; moreover, it is a function of a society comprising motivated and in some sense 'included' people. Bringing about such a society needs to be worked on by 'brave' government, as do the extension, refinement and incorporation of intermediary participatory tools and techniques (cf. LGMB, 1999; Clarke and Stewart, 1998; Forester, 1999).

There are two further things to be noted in this regard. First, many people do feel disenfranchised and do not participate in formal political modes, and are probably not engaged in the types of actions detailed in the examples given. Second, there are others who use a range of mechanisms that post-modern politics allows in order to make political capital. It is difficult for the public and for governments to discern what case or claim is moral or justifiable. It is the question of different sources, mediators and exercises of power that appears key. The way that such resources are used and abused in rural affairs requires careful attention by researchers and politicians alike. Klug (1997), in discussing rights and policy in the UK, has suggested that 'rights impact assessments' could be incorporated into policy-making. Although not entirely the same thing, the concept of policy 'proofing' has begun to gain currency in Whitehall. For example, the PIU report *Rural Economies* (PIU, 1999) and the 2000 rural White Paper (DETR, 2000a) incorporate this idea whereby policy proposals across government departments are assessed in terms of the potential impact that they hold for rural areas (primarily in terms of impact on the rural economy). The idea of rights assessment takes the rural proofing notion further, whereby policy should be assessed in terms of impact on citizenship and rights/responsibilities – in one sense, a form of social visioning. Thus, if Klug's idea were adapted to become a wider notion of 'citizenship assessment', it could provide a framework for groups, sociations or communities to work out how particular policies or other drivers of change might impact on them and on others. Such a framework appears a logical extension of parish and village appraisals. An obstacle is the likely partiality of such participation and the cost of extensifying such mechanisms to ensure inclusivity. Nevertheless, if citizenship is processual, then the lead and animation of citizenship should similarly be empowering in the sense of enabling independence for particular 'information-poor' communities or otherwise marginalised groups and allowing a consequent widening of concern. Such empowering also takes the emphasis on the perpetuated urban/rural divide away from the process, otherwise why not 'urban-proof' policy proposals?

Attempts to inform and prepare citizens (e.g. Day, 1998) are still essentially tools to enable reaction rather than prepare action in the proactive or positive sense. Some approaches and planning tools attempt to engage people in future-gazing – for example, LAZI21, parish and village appraisals (Moseley *et al.*, 1996; Countryside Agency, 1999a; DETR, 2000a) – but still these have tended to rely on already educated and 'active' middle-class participants. There seems to be an incredible 'skills gap' that partners the democratic and citizenship deficit. This is compounded by the disillusionment of many towards established governance practice and procedure by others who are active in other ways. Another area to explore therefore is the more practical and education-regarding political systems and decision-making. These ought to produce an increasingly informed and critical citizenry and to help people be more vociferous and argumentative. However, improving levels of education are not being matched by any increase in engagement with the formal, conventional political process and systems of decision-making. While *citizenship* is to become a compulsory part of the English and Welsh National Curriculum from 2002, what kind of citizenship is to be taught? As this examination shows, a great deal of care should be taken over such a subject. But it is clear that a citizen-facing system of governance needs to be engendered, although it is beyond the ambit of this work to proselytise further on that. However, deeper and wider participation might involve 'going and getting' structured citizenship engagement on the part of government, local government and other institutions.

Through the course of the text there has been an implied and often explicit criticism of models and policies of citizenship and associated use of citizenship rhetoric. Attention has focused on those efforts and alternative practices where individuals and groups have looked to engage with authority in what might be termed critical active citizenship. This postmodern reading, however, does require a further issue to be addressed: how can citizens more effectively engage with political processes at different scales and in different spheres (and here, as they relate to the countryside)?

This reading of citizenship and countryside politics has allowed for a consideration that highlights the constructed nature of citizenship that is evolving as a result of economic change at the global level and political change at the local and national scale. It is argued that managing change effectively requires decision-making that is *deliberative* (i.e. carefully considered) and *inclusionary* (i.e. including as wide a range of interested parties as possible; see Forester, 1999). However, it must be demonstrated that the quality of the final outcome improves, yet how can this be measured? In the Wye Valley example it was possible for a strong individual to gather resources and present challenges to authority. This type of engagement is rare and undertaken by a tiny minority. At least that and other examples cited demonstrate (contra to cultural commentators such as Adorno, 1990) that regulation and consumerism, and the associated rise of the culture industry, can provide new forms of resistance as well as (putative) domination by the market and associated commodity relations. In at least one sense, all protest is healthy in forcing public and inward scrutiny of decision-makers and decision-making.

Deliberative and inclusionary techniques may help improve the process of decision-making and actually impact on policy, and they can get a range of people involved, although in this respect the notion of the stakeholder does require more careful consideration in future scholarship. Interconnectedness and cultural diversity imply a range of potential sub-interests to be considered. Focus groups have been critiqued recently as they form an important part of Tony Blair's approach to gauging public opinion in the UK. This is interesting and shows how such techniques appear to be considered good enough to help shape policy at the national level – particularly where intro-duced informally. To be cynical, perhaps they are used because the findings can be kept private and/or ignored. Such approaches should be reflected more in research methodologies that aim to uncover the complex cultural and social relations and imaginations that appear to underpin our use of and policies towards the countryside and to particular rural industries, or other facets of countryside policy. Countryside policy might well require a more devolved regional power-share and ensuring that particular groups and issues are debated regularly. Is it not healthy to revisit issues and ask if we are still sure that we are doing things right? In terms of land policy this might be usefully done in terms of planning, agriculture and the whole range of related issues that arise from those facets of public/private regulation. In essence, qualitative methods should be encouraged both by researchers and citizens themselves in order to investigate and present rural issues and opinions, problems and predicaments.

Rural studies, land and citizenship

In the UK context it is contended that we should be exploring more radical and forward-looking options when it comes to citizenship, and particularly how we regulate (legally and culturally) each other and each other's use of the land. Looking for such options may require a carefully restructured plan-ning system and (specific to the concerns of this text) a determination on the part of government to control land value and land use imaginatively and flexibly. Questions of land reform and policy are prompted by the need to reorganise production (and even land values) and for ecological/environ-mental motives. It is argued that land reform should be viewed as a key component of the restructuring of institutions and other structures in post-national, post-industrial society (even to the extent of ensuring that transaction costs and bureaucracy are minimised). This issue, and the related brokerage of property rights, should be brought forward again for public discussion in England and Wales, especially given the recent partial reforms introduced in Scotland (Scottish Executive, 1999), as well as the incorpo-ration of the Human Rights Act (1998), which took effect there prior to its commencement in England. Another reason why a re-examination of land reform is needed is in the light of environmental priorities, given the importance of land to both society and ecology. This platform of change and redistribution could also provide a springboard to look at the role of the welfare state and relationships to natural resources, including how a vibrant, diverse countryside polis might be better enabled.

An attempt has been made to introduce a different way of looking at and using citizenship, as well as applying such a frame to rural/environment studies. It is also convenient that citizenship theory provides a synthetic approach to tie history and the contemporary and different theoretical and methodological approaches towards research into citizenships, contingency and the countryside. In terms of matters that do relate more directly to rural studies it seems clear that the approach could be extended, refined and applied to many other contexts or policy areas. Attention could be turned to other case studies in local empowerment or 'active' citizenship in the context of the rural and the environment. Such concerns as the study of how the 'everyday' (see Crouch, 1997, 2000; Shotter, 1993; De Certeau, 1984) is constitutive of citizenship and the creation of rights and responsibilities (of all types) through cultural practices would certainly be welcome. Similarly, the impact of information of different types, including 'technological citizenship', should be investigated.

The last three chapters provide a diverse and yet partial indication of how a widened conceptualisation of citizenship might be viewed and investigated. They show that citizen action is diverse and contextual, often led by the media or by NIMBYist forms of self-interest. One of the strengths of citizen action is that it requires and should engender critical and proactive engagement with issues. The contestation of rights and the exercise of diverse strategies by groups and individuals are out there to be found. The examples used cover a range of actions and reaction, but provide a far from complete picture. Further study can utilise the citizenship frame to good effect, showing how and why people take part in particular practices and what effects their actions have. How they go about challenging or subverting extant legal or cultural citizenship envelopes, as well as day-to-day or normalised practices that support existing power relations (or only gradually alter them), is a very important project.

As detailed, citizenship may be viewed as the practice and exchange of power between competing communities, groups and individuals. The way that legal and cultural rights are traded or contested as part of the process of multilevel and cross-cutting politics, based on diverse binding and separating characteristics and imaginations, has been identified as a key area for research and one that can bring together previously disparate areas or topics of study. It widens and links citizenship with the brokerage of power. Hill (1974: 157) notes the connection that 'the argument for greater citizen participation is an argument about power' and, in connection with rural and environmental policy and planning, 'power is the crucial issue; who is to decide local policy and where control is to lie, are central' (*ibid.*). Thus citizenship, participation and decision-making should be linked explicitly and clearly to engender a culture of 'making a difference', partly because it is patently the case that people can and do create change: both structured and diffuse, proactively and reactively.

Our understanding of citizenship is widened as it becomes redefined. Citizenship is fluid and its importance is political, cultural and economic. It also has a sign value that is appropriated by groups and interests at all scales from the local to the global. Citizenship as a result has become a discursive vehicle for both change and resistance to change. This implies that citizenship action moves back towards the political and the social rather than being dominated by the cultural and the economic. In the global age, fragments of time, culture and history are appropriated to support political contests, and 'storytelling' using rights, obligations and moral arguments based on past actions is variously deployed. It may be argued that (the traditional conception of) citizenship itself is primarily a signification: a hyper-real fragment of modernity, and the multiple, mirrored versions presented by protagonists and exhibited by others in postmodern politics are a fractured reflection of this. Citizenship in and of the countryside is constructed, appropriated and contingent: and it is now more than ever, clearly beyond the control of the nation-state.

Bibliography

Abercrombie, N., Hill, S. and Turner, B. (1980) *The Dominant Ideology Thesis*. Allen and Unwin, London.

Abrams, S. (2000) 'Planning the Public: Some Comments on Empirical Problems for Planning Theory'. *Journal of Planning Education and Research* 19: 351–357.

Ackerman, P. and Alstott, J. (1999) *The Stakeholder Society*. Harvard University Press, Cambridge, Mass.

Adam, B. (1995) *Timewatch: A Social Analysis of Time*. Polity, Cambridge.

Adorno, T. (1990) 'Culture Industry Reconsidered', in Alexander, J. and Seidman, S. (eds) *Culture and Society. Contemporary Debates*. Cambridge University Press, Cambridge.

Agyeman, J. and Kinsman, P. (1997) 'Analysing macro- and micro-environments from a multi-cultural perspective', in Hooper-Greenhill, E. (ed.) *Cultural Diversity: Developing Museum Audiences in Britain*. Leicester University Press, London.

Albrow, M. (1996) *The Global Age*. Polity, Cambridge.

Alinski, S. (1972) *Rules for Radicals: A Practical Primer*. Random House, New York.

Allison, L. (1975) *Environmental Planning*. Allen and Unwin, London.

Allmendinger, P. (2001) *Planning in Postmodern Times*, Routledge, London.

Ambrose, P. (1986) *Whatever Happened to Planning?* Methuen, London.

Anderson, A. (1997) *Media, Culture and the Environment*. UCL Press, London.

Anderson, B. (1983) *Imagined Communities*. Verso, London.

Anfield, J. (2001) 'A Real Opportunity for Rural Planners'. *Town and Country Planning*, January: 19.

Appadurai, A. (1990) 'Disjuncture and Difference in the Global Cultural Economy'. *Theory, Culture and Society* 7: 296–310.

Archer, M. (1988) *Culture and Agency: The Place of Culture in Social Theory*. Cambridge University Press, Cambridge.

Arendt, H. (1977) *Crises of the Republic*. Penguin, Harmondsworth.

Arnstein, S. (1971) 'A Ladder of Citizen Participation'. *Journal of the American Institute of Planners* 35 (4): 216–224.

Barnett, A. and Scruton, R. (eds) (1999) *Town and Country*. Vintage, London.

Barrow, C. (1997) *Environmental and Social Impact Assessment*. Arnold, London.

Batie, S. (1984) 'Alternative Views of Property Rights: Implications for Agricultural Use of Natural Resources'. *American Journal of Agricultural Economics*, December: 814–818.

Baudrillard, J. (1988) *For a Critique of the Political Economy of the Sign*. Telos, St Louis.

Baudrillard, J. (1993) *Symbolic Exchange and Death*. Sage, London.
Baudrillard, J. (1998) *The Consumer Society*. Sage, London. (First published 1970.)
Bauman, Z. (1987) *Legislators and Interpreters*. Polity, Oxford.
Bauman, Z. (1992) *Intimations of Postmodernity*. Routledge, London.
Bauman, Z. (1998) *Globalisation: The Human Consequences*. Polity, Cambridge.
Bauman, Z. (2000) *Liquid Modernity*. Polity, Cambridge.
BBC TV (1998) *Sparring Partners*. Documentary on the Brent Spar controversy. BBC Television, London.
BBC TV (1999) 'In for the Kill'. *Blood on the Carpet* documentary series. BBC Television, London.
Bebbington, S. (1999) Press release issued by Simon Bebbington headed 'Diggers Squat Des Res Land', 4 April.
Beck, U. (1992) *Risk Society*. Sage, London.
Beck, U. (1998) 'Politics of Risk Society', in Franklin, J. (ed.) *The Politics of Risk Society*. Polity, Cambridge.
Becker, L. (1977) *Property Rights: Philosophic Foundations*. Routledge, London.
Beckett, J. (1990) *The Agricultural Revolution*. Blackwell, Oxford.
Beer, J. (1997) 'My Ace in the Hole Has Been Trumped', *Daily Telegraph*, 8 February.
Bell, D. and Valentine, G. (1997) *Consuming Geographies*. Routledge, London.
Bender, B. (1993) 'Stonehenge: Contested Landscapes (Medieval to Present-Day)', in Bender, B. (ed.) *Landscape: Politics and Perspectives*. Berg, Providence/Oxford.
Berger, J. (1973) *Ways of Seeing*. Penguin, Harmondsworth.
Berman, M. (1983) *All That's Solid Melts to Air*. Verso, London.
Bey, H. (1991) *T.A.Z.: The Temporary Autonomous Zone, Ontological Anarchy, Poetic Terrorism*, Autonomedia, Brooklyn, NY.
Bijker, W. and Law, J. (1992) *Shaping Technology/Building Society*. MIT Press, Cambridge, Mass.
Bijker, W., Hughes, T. and Pinch, T. (1989) *The Social Construction of Technological Systems*. MIT Press, Cambridge, Mass.
Billinge, M. (1993) 'Trading History, Reclaiming the Past: The Crystal Palace as Icon', in Kearns, G. and Philo, C. (eds) *Selling Places: City as Cultural Capital Past, Present and Future*. Pergamon, Oxford.
Birchall, J. (1994) *Co-Op: The People's Business*. Manchester University Press, Manchester.
Bird, J., Curtis, B., Putnam, T., Robertson, G. and Tickner, L. (eds) (1993) *Mapping the Futures: Local Cultures, Global Change*. Routledge, London.
Bishop, K. (1998) 'Countryside Conservation and the New Right', in Allmendinger, P. and Thomas, H. (eds) *Urban Planning and the British New Right*. Routledge, London.
Bishop, K. and Flynn, A. (1999) 'The National Assembly for Wales'. *Countryside Recreation* 7 (Winter): 5–8.
Blackburn, R. (1997) 'A British Bill of Rights for the 21st Century', in Blackburn, R. and Busuttil, J. (eds) *Human Rights in the 21st Century*. Pinter, London.
Blackman, D. (2001) 'Rights and Wrongs'. *Property Week*, 12 January: 72–75.
Blair, T. (1997) *Text of Conference Speech*. Delivered at the 1997 Annual Labour Party conference, Brighton, September. Labour Party, London.
Blair, T. (1999) *Text of Conference Speech*. Delivered at the 1999 Annual Labour Party conference, Brighton, September. Labour Party, London.

Blair, T. (1999) *Text of Conference Speech*. Delivered at the 1999 Annual Labour Party conference, Brighton, September. Labour Party, London.

Blomley, N. (1989) 'Interpretive Practices: The State and the Locale', in Wolch, J. and Dear, M. (eds) *The Power of Geography*. Unwin Hyman, Boston.

Blomley, N. (1994) 'Mobility, Empowerment and the Rights Revolution'. *Political Geography* 13 (5): 407–422.

Blomley, N. (1995) *Law, Space and the Geographies of Power*. Guilford Press, London.

Bloomfield, D., Collins, K., Fry, C. and Munton, R. (1998) 'Deliberative and Inclusionary Processes: Their Contribution to Environmental Governance'. Paper given at the Deliberative and Inclusionary Processes seminar, UCL, London, 17th December 1998.

Blowers, A. (ed.) (1993) *Planning for a Sustainable Environment*. Earthscan, London.

Blunden, J. and Curry, N. (eds) (1990) *A People's Charter?* HMSO, London.

Boettger, O. (1998) 'New Actor-Network Resource'. Website at <http://www.mailbase.ac.uk/hypermail/lists-p-t/sosig/1997–04/0012.html>.

Bonnes, M. and Secchiaroli, G. (1995) *Environmental Psychology*. Sage, London.

Bonyhady, T. (1987) *The Law of the Countryside*. Professional Books, Oxford.

Boulton, D. (1999) *Gerrard Winstanley and the Republic of Heaven*. Dales Historical Monographs, Dent, Cumbria.

Bourdieu, P. (1977) *Outline of a Theory of Practice*. Cambridge University Press, Cambridge.

Bourdieu, P. (1984) *Distinction: A Social Critique of the Judgement of Taste*. Routledge, London.

Bourdieu, P. (1990) *The Logic of Practice*. Polity Press, Cambridge.

Bourdieu, P. (1991) *Language and Symbolic Power*. Polity, Cambridge.

Bourdieu, P. (1993) 'The Field of Cultural Production', in *The Polity Cultural Studies Reader*. Polity, Cambridge.

Bourdieu, P. and Passeron, J.-C. (1977) *Reproduction in Education, Society and Culture*. Sage, London.

Bowers, J. and Cheshire, P. (1983) *Agriculture, the Countryside and Land Use*. Methuen, London.

Boyle, P. and Halfacree, K. (1998) *Migration into Rural Areas: Theories and Issues*. Wiley, Chichester.

Bradstock, A. (1997) *Faith in the Revolution*. SPCK, London.

Bradstock, A. (ed.) (2001) *Winstanley and the Diggers*. Frank Cass, London.

Briggs, A. (1961) *Social Thought and Social Action*. Allen and Unwin, London.

Brindley, T., Rydin, Y. and Stoker, G. (1989) *Remaking Planning*. Unwin Hyman, London.

Brindley, T. Rydin, Y. and Stoker, G. (1996) *Remaking Planning: The Politics Of Urban Change* 2nd edition. Routledge, London.

Bromley, D. (1991) *Environment and Economy: Property Rights and Public Policy*. Blackwell, Oxford.

Bromley, D. (1998) 'Rousseau's Revenge: The Demise of the Freehold Estate', in Jacobs, H. (ed.) *Who Owns America?* University of Wisconsin Press, Madison.

Bromley, D. and Hodge, I. (1990) 'Private Property Rights and Presumptive Entitlements: Reconsidering the Premises of Rural Policy'. *European Review of Agricultural Economics* 17: 197–214.

Brown, P. (1996a) 'Boys from the Green Stuff'. *Guardian Society*, 29 May: 5.

Brown, P. (1996b) 'Tarmac Call for Greener New Roads'. *Guardian*, 21 May: 2.

Bucke, T. and James, Z. (1998) *The Impact of the Criminal Justice and Public Order Act (1994)*. Home Office, London.

Buckingham-Hatfield, S. and Percy, S. (1998) *Constructing Local Environmental Agendas*. Routledge, London.

Bulmer, M. and Rees, A. (eds) (1996) *Citizenship Today*. UCL Press, London.

Bunce, M. (1994) *The Countryside Ideal*. Routledge, London.

Burgess, J. and Harrison, C. (1998) 'Environmental Communication and the Cultural Politics of Environmental Citizenship'. *Environment and Planning A* 30: 1445–1460.

Burningham, K. (2000) 'Using the Language of NIMBYism: A Topic for Research, not an Activity for Researchers'. *Local Environment* 5 (1): 55–67.

Burns, D., Hambleton, R. and Hoggett, P. (1994) *The Politics of Decentralisation*. Macmillan, Basingstoke.

Burns Committee (2000) Website of the Committee of Inquiry into Hunting with Dogs at <www.huntinginquiry.gov.uk> (accessed 24 August 2000).

Callon, M. (1980) 'Struggles and Negotiations to Define What Is Problematic and What Is Not: The Sociologic Translation', in Knorr, K., Krohn, R. and Whitley, R. (eds) *The Social Process of Scientific Investigation*. Reidel, Dordrecht.

Callon, M. (1986) 'Some Elements of a Sociology of Translation: Domestication of the Scallop Fishermen of St. Brieuc Bay', in Law, J. (ed.) *Power, Action, Belief: A New Sociology of Knowledge*. Routledge and Kegan Paul, London.

Callon, M. (1991) 'Techno-economic Networks and Irreversibility', in Law, J. (ed.) *A Sociology of Monsters*. Routledge, London.

Callon, M. and Latour, B. (1981) 'Unscrewing the Big Leviathan', in Knorr-Cetina, K. and Cicourel, A. (eds) *Advances in Social Theory and Methodology*. Routledge and Kegan Paul, London.

Card, R. and Ward, R. (1996) 'Access to the Countryside: The Impact of the Criminal Justice and Public Order Act 1994'. *Journal of Environment and Planning Law*, June: 447–462.

Carson, R. (1962) *Silent Spring*. Penguin, Harmondsworth.

Castells, M. (1997) *The Rise of the Network Society*, vol. 1. Macmillan, Basingstoke.

Chambers, J. and Mingay, G. (1966) *The Agricultural Revolution 1750–1880*. Batsford, London.

Charlesworth, A. (1980) 'The Development of the English Rural Proletariat and Social Protest 1700–1850: A Comment'. *Journal of Peasant Studies* 8 (1): 101–111.

Charter88 (2000) Website at <www.charter88.org.uk/citizenship> (accessed 15 September 2000).

Cherry, G. (1973) *Town Planning in Its Social Context*. Leonard Hill, Aylesbury.

Cherry, G. (1975) *History of Environmental Planning*, vol. 2. HMSO, London.

Cherry, G. and Rogers, A. (1996) *Rural Change and Planning*. E. and F. N. Spon, London.

Chesters, G. (2000) 'Guerrilla Gardening: The End of the World as We Know It?' *Ecos* 21 (1): 10–13.

Cigler, A. (1991) 'Interest Groups: A Subfield in Search of an Identity', in Crotty, W. (ed.) *Political Science: Looking to the Future*, vol. 4. Northwestern University Press, Evanston, Ill.

Citizens' Income Trust (2000) Trust website at <www.ownbase.org.uk/citizensincome> (accessed March 2000).

Clark, G., Darrall, J., Grove-White, R., Macnaghten, P. and Urry, J. (1994) *Leisure Landscapes* (Main Report). Council for the Protection of Rural England, London.

Clarke, J. and Critcher, C. (1985) *The Devil Makes Work*. Macmillan, Basingstoke.

Clarke, M. and Stewart, J. (1998) *Modernising Local Government*. INLOGOV, Birmingham.

Clarke, P. (ed.) (1994) *Citizenship: A Reader*. Pluto Press, London.

Clarke, P. (1996) *Deep Citizenship*. Pluto Press, London.

Clawson, M. and Knetsch, J. (1965) *Economics of Outdoor Recreation*. Johns Hopkins University Press, Baltimore.

Clayden, P. (1992) *Our Common Land, the Law and History of Commons and Village Greens*. OSS, Henley.

Clegg, S. (1989) *Frameworks of Power*. Sage, London.

Clegg, S. (1994) 'Power Relations and the Constitution of the Resistant Subject', in Jermier, J., Knights, D. and Nord, W. (eds) *Resistance and Power in and around Organisations*. Routledge, London.

Cloke, D. (1996) 'Critical Writing in Rural Studies'. *Sociologica Ruralis* 36 (1): 117–119.

Cloke, P. and Little, J. (eds) (1997) *Contested Countryside Cultures*. Routledge, London.

Cloke, P., Doel, M., Matless, D., Philips, M. and Thrift, N. (1995) *Writing the Rural*. Paul Chapman, London.

Cloke, P., Goodwin, M. and Milbourne, P. (1998) 'Inside Looking Out; Outside Looking In: Different Experiences of Cultural Competence in Rural Lifestyles', in Boyle, P. and Halfacree, K. (eds) *Migration into Rural Areas: Theories and Issues*. Wiley, Chichester.

Cloke, P., Milbourne, P. and Widdowfield, R. (2000) 'Homelessness and Rurality: "Out of Place" in Purified Space?'. *Environment and Planning D: Society and Space* 18: 715–735.

Cohen, A. (1989) *The Symbolic Construction of Community*. Routledge, London.

Cohen, S. (ed.) (1972) Introduction to *Images of Deviance*. Penguin, Harmondsworth.

Cooper, D. (1993) 'The Citizen's Charter, and Radical Democracy: Empowerment and Exclusion within Citizenship Discourse'. *Social and Legal Studies* 2: 149–171.

Cooper, D. (1998) *Governing out of Order*. Rivers Oram Press, London.

Coote, A. (1998) 'Risk and Public Policy: Towards a High-Trust Democracy', in Franklin, J. (ed.) *The Politics of Risk Society*. Polity, Cambridge.

Cornwall Society, The (2000) Website at <www.cornwall.eu.org/assembly> (accessed 12 August 2000).

Costain Group (1997a) Personal communication. Interview with Director of Communications, Costain Group, 3 October.

Costain Group (1997b) *Costain and the Environment: A Working Partnership*. The Costain Group PLC, London.

Country Landowners Association (1991) *A Better Way Forward*. CLA, London.

Country Landowners Association (1996) *Access 2000*. CLA, London.

Country Landowners Association (1998) 'Country Landowners Association Response to the Government Consultation Paper on Access to the Countryside', June. CLA, London.

Countryside Agency (1999a) *Planning for the Countryside*. Countryside Agency, Cheltenham.

Countryside Agency (1999b) *The State of the Countryside 1999.* Countryside Agency, Cheltenham.

Countryside Agency (2000a*) The Local Heritage Initiative.* Countryside Focus no. 6. Countryside Agency, Cheltenham.

Countryside Agency (2000b) 'Local Food, National Support'. Press release 3 July, accessed 8 September 2000 on the Countryside Agency website <www.countryside.gov.uk/news/article>.

Countryside Agency (2000c) 'Scheme Puts Heritage in Hands of Locals'. *Countryside Focus* no. 6 (February).

Countryside Agency (2001) *New Rights, New Responsibilities: What the New Countryside Access Arrangements Will Mean to You.* Countryside Agency, Cheltenham.

Countryside Alliance (1998) Countryside March flyer, March. Countryside Alliance, London.

Countryside Alliance (1999) *Statement of Aims.* Website at <www.countrysidealliance.org> (accessed July 1999).

Countryside Alliance (2000) *Manifesto.* At <www.countrysidealliance.org/alliance/brochure.htm> (accessed 18 September 2000).

Countryside Commission (1987) *Enjoying the Countryside: Priorities for Action.* The Commission, Cheltenham.

Countryside Commission (1989) *Managing Rights of Way: An Agenda for Action.* The Commission, Cheltenham.

Countryside Commission (1993a) *Annual Rights of Way Expenditure Survey 1990–1.* The Commission, Cheltenham.

Countryside Commission (1993b) *National Target for Rights of Way: A Guide to Milestones.* The Commission, Cheltenham.

Countryside Commission (1994) *Parish Paths Partnership: An Outline.* The Commission, Cheltenham.

Countryside Commission (1996a) *Priorities for Action: The Next Ten Years.* Countryside Commission, Cheltenham.

Countryside Commission (1996b) *A Vision for the Next Ten Years.* The Commission, Cheltenham.

Countryside Commission (1998) *Planning for Countryside Quality.* Policy statement. Countryside Commission, Cheltenham.

Countryside Movement (1995) *Our Charter.* Countryside Movement, London.

Cowe, R. (1999) 'Firms Want Help to Be Responsible'. *Guardian*, 8 November: 8.

Cox, A. (1984) *Adversary Politics and Land.* Cambridge University Press, Cambridge.

Cox, G. and Winter, M. (1997) 'The Beleaguered "Other": Hunt Followers in the Countryside', in Milbourne, P. (ed.) *Revealing Rural Others.* Pinter, London.

Cox, G., Lowe, P. and Winter, M. (1990) *The Voluntary Principle in Conservation.* Packard, Chichester.

Crang, M. (1997) 'Picturing Practices: Research through the Tourist Gaze'. *Progress in Human Geography* 21 (3): 359–373.

Crang, M. (1998) *Cultural Geography.* Routledge, London.

Cresswell, T. (1993) 'Mobility as Resistance: A Geographical Reading of Kerouac's "On the Road"'. *Transactions of the Institute of British Geographers*, NS 18: 249–262.

Cresswell, T. (1996) *In Place/Out of Place: Geography, Ideology and Transgression.* University of Minnesota Press, Minneapolis.

Crouch, D. (1992) 'Popular Culture and What We Make of the Rural'. *Journal of Rural Studies* 8: 229–240.

Crouch, D. (1997) '"Others" in the Rural: Leisure Practices and Geographical Knowledge', in Milbourne, P. (ed.) *Revealing Rural Others*. Pinter, London.

Crouch, D. (1999) 'Introduction: Encounters in Leisure/Tourism', in Crouch, D. (ed.) *Leisure/Tourism Geographies*. Routledge, London.

Crouch, D. (2000) 'Places around Us: Embodied Lay Geographies in Leisure and Tourism'. *Leisure Studies* 19: 63–76.

Crouch, D. and Matless, D. (1996) 'Refiguring Geography: The Parish Maps of Common Ground'. *Transactions of the IBG Institute of British Geographers*, NS 21 (1): 236–255.

Crouch, D. and Parker, G. (forthcoming) '"Digging Up Heritage" and Diffuse Politics: (Re)politicising History by Representing Place and Practice'.

Crouch, D. and Ravenscroft, N. (1995) 'Culture, Social Difference and the Leisure Experience', in McFee, G. (ed.) *Leisure Cultures*. LSA, Brighton.

Crow, G. (1996) 'Taking Stock of Recent Rural Studies in the UK: A Reply to Miller and Cloke'. *Sociologica Ruralis* 36 (3): 361–364.

Crow, G. and Allan, G. (1996) *Community Life*. Harvester Wheatsheaf, Hemel Hempstead.

Cullingworth, J. (1994) 'Fifty Years of Post-war Planning'. *Town Planning Review* 65 (3): 277–304.

Cullingworth, J. (1999) *Fifty Years of Urban and Regional Policy*. Athlone Press, London.

Cullingworth, J. and Nadin, V. (1997) *Town and Country Planning in the UK*. Routledge, London.

Curry, N. (1993) 'Countryside Planning: Look Back in Anguish'. Inaugural Lecture, Cheltenham, 28 April 1993. Cheltenham and Gloucester College of Higher Education, Cheltenham.

Curry, N. (1994) *Countryside Recreation: Access and Land Use Planning*. E. and F. N. Spon, London.

Curry, N. (1996) 'Access: Policy Directions for the Late 1990s', in Watkins, C. (ed.) *Rights of Way: Policy, Culture and Management*. Pinter, London.

Curry, N. and Ravenscroft, N. (1996) 'Charging for Public Path Orders'. Unpublished report for the Department of the Environment by CCRU, Cheltenham.

Dahrendorf, R. (1979) *Life Chances*. University of Chicago Press, Chicago.

Dahrendorf, R. (1994) 'The Changing Quality of Citizenship', in Van Steenbergen, B. (ed.) *The Condition of Citizenship*. Sage, London.

Darke, R. (1999) 'Public Speaking Rights in Local Authority Planning Committees'. *Planning Practice and Research* 14 (2): 171–183.

Davies, A. (1999) 'Where Do We Go from Here? Environmental Focus Groups and Planning Policy Formation'. *Local Environment* 4 (93): 295–316.

Day, M. (ed.) (1998) *Environmental Action: A Citizens' Guide*. Pluto Press, London.

Day, P. (1995) *Land: The Elusive Quest for Social Justice, Taxation Reform and Global Environmental Sustainability*. Australian Academic Press, Brisbane.

De Certeau, M. (1984) *The Practice of Everyday Life*. University of California Press, London.

Denman, D. (1978) *The Place of Property*. Geo Press, Berkhamstead.

Department of Transport (1989) *Roads for Prosperity*. Transport White Paper. DoT/HMSO, London.

DETR (Department of the Environment, Transport and the Regions) (1997) *Local Government and Rating Act 1997: Parish Reviews.* Environment Circular 11/97, July. Department of the Environment, Transport and the Regions, London.

DETR (1998a) *Modernising Local Government.* Consultation paper, February. Department of the Environment, Transport and the Regions, London.

DETR (1998b) *Access to the Open Countryside in England and Wales.* Consultation paper, February. Department of the Environment, Transport and the Regions, London.

DETR (1998c) *Modernising Local Government: Local Democracy and Community Leadership.* Consultation document, February. Department of the Environment, Transport and the Regions, London.

DETR (1999) *Access to the Countryside: A Consultation.* November. Department of the Environment, Transport and the Regions, London.

DETR (2000a) *Our Countryside: The Future.* The Rural White Paper, November. Department of the Environment, Transport and the Regions, London.

DETR (2000b) *Our Towns and Cities: The Future.* The Urban White Paper, November. Department of the Environment, Transport and the Regions, London.

Dickson, L. and McCulloch, A. (1996) 'Shell, Brent Spar and Greenpeace: A Doomed Tryst?'. *Environmental Politics* 5 (1): 122–129.

Dodgshon, R. (1999) 'Human Geography at the End of Time? Some Thoughts on the Notion of Time–Space Compression'. *Environment and Planning D: Society and Space* 17: 607–620.

DoE/MAFF (Department of the Environment/Ministry of Agriculture, Fisheries and Food) (1995) *Rural England: A Nation Committed to a Living Countryside.* White Paper, November. Department of the Environment/Ministry of Agriculture, Fisheries and Food, London.

Donnelly, P. (1986) 'The Paradox of Parks: Politics of Recreational Land Use'. *Leisure Studies* 5 (2): 211–231.

Donnelly, P. (1994) 'The Right to Wander: Issues in the Leisure Use of Countryside and Wilderness Areas'. *Leisure Studies* 28: 187–201.

Donzelot, J. (1980) *The Policing of Families.* Hutchinson, London.

Douglas, R. (1976) *Land, People and Politics: A History of the Land Question in the United Kingdom, 1878–1952.* Allison and Busby, London.

Dugdale, A. (1999) 'Materiality: Juggling Sameness and Difference', in Law, J. and Hassard, M. (eds) *Actor Network Theory and After.* Blackwell, Oxford.

Dump the Pump (2000) Website at <www.dumpthepump.co.uk>.

Eder, K. (1993) *The New Politics of Class: Social Movements and Cultural Dynamics in Advanced Societies.* Sage, London.

Edwards, W., Pemberton, M. and Woods, M. (1999) 'Governing Governance . . .'. Paper given at the European Rural Sociology Conference, Lund, Sweden, August 1999.

Elias, N. (1982) *The Civilising Process*, vol. 2: *State Formation and Civilisation.* Blackwell, Oxford.

Engel, M. (1998) 'The Day the City Became a Shire'. *Guardian*, 2 March: 1.

Environment Agency (1996) *Quays for Orimulsion Unloading at Milford Haven.* Press statement, July. Environment Agency, Cardiff.

Environment Agency (1997) Deposition to the Wye Navigation Order Public Inquiry March 1997. Document ref. EA/E1. Environment Agency, Monmouth.

Environment Agency (1998) *Local Environment Agency Plan: Wye Area Consultation Report*. November. Environment Agency Wales, Cardiff.

Environment Agency (2001) Personal communication from Anthony Weare, Environment Agency solicitor, Cardiff.

Environment Council (1995) *Beyond Compromise: Building Consensus in Environmental Planning and Decision Making*. Environmental Resolve. Environment Council, London.

Environmental Resolve (1995) *Beyond Compromise: Building Consensus in Environmental Planning*. Environment Council, London.

Etzioni, A. (1993) *The Spirit of Community*. Touchstone, New York.

Etzioni, A. (2000) 'You Have Fixed the Economy, Mr Blair – Now You Must Mend Society'. *The Times*, 5 July.

Fagence, M. (1985) *Citizen Participation in Planning*. Pergamon, Oxford.

Fairlie, S. (1996) *Low Impact Development*. Jon Carpenter, Charlbury.

Fairlie, S. (2000) 'If Property Is Theft Is Planning a Diversion?'. *Chapter 7 News* no. 4 (Summer).

Falk, I. and Kilpatrick, S. (2000) 'What Is Social Capital? A Study of Interaction in a Rural Community'. *Sociologia Ruralis* 40 (1): 87–110.

Falk, R. (1994) 'The Making of Global Citizenship', in Van Steenbergen, B. (ed.) *The Condition of Citizenship*. Sage, London.

Falk, R. (2000) 'The Decline of Citizenship in an Era of Globalization'. *Citizenship Studies* 4 (1): 5–17.

Featherstone, M. (1991) *Consumer Culture and Postmodernism*. Sage, London.

Fine, B. (2001) *Social Capital versus Social Theory*. Routledge, London.

Fisher, J. (2000) 'Property Rights in Pheasants: Landlords, Farmers and the Game Laws 1860–80'. *Rural History* 11 (2): 165–180.

Flyvbjerg, B. (1998) *Rationality and Power*. University of Chicago Press, Chicago and London.

FoE (Friends of the Earth) (1997) Personal communication regarding shareholder protest, 1 October. Friends of the Earth, London.

Forester, J. (1999) *The Deliberative Practitioner*. MIT Press, Cambridge, Mass.

Foucault, M. (1977) *Discipline and Punish*. Penguin, Harmondsworth.

Foucault, M. (1980a) *Power/Knowledge: Selected Interviews*, ed. C. Gordon. Harvester, Brighton.

Foucault, M. (1980b) 'Powers and Strategies', in *Power/Knowledge: Selected Interviews*, ed. C. Gordon. Harvester Wheatsheaf, Brighton.

Foucault, M. (1980c) *An Archaeology of Knowledge*. Harvester, Brighton.

Foucault, M. (1988) *Politics, Philosophy, Culture: Interviews and Other Writings 1977–1984*. Routledge, Chapman and Hall, New York.

Franklin, J. (ed.) (1998) *The Politics of Risk Society*. Polity, Cambridge.

Fraser, N. and Gordon, L. (1994) 'Civil Citizenship versus Social Citizenship', in Van Steenbergen, B. (ed.) *The Condition of Citizenship*. Sage, London.

Fudge, C. and Glasbeek, H. (1992) 'The Politics of Rights: A Politics with Little Class'. *Social and Legal Studies* 1: 45–70.

Fukayama, F. (1992) *The End of History and the Last Man*. Penguin, London.

Fyfe, N. (1995) 'Law and Order Policy and Spaces of Citizenship in Contemporary Britain'. *Political Geography* 14 (2): 177–189.

Gadsden, G. (1987) *Law of the Commons*. Sweet and Maxwell, London.

Gamberale, C. (1997) 'European Citizenship and Political Identity'. *Space and Polity* 1 (1): 37–59.

Gamble, A. (1988) 'The Politics of Thatcherism'. *Parliamentary Affairs* 42 (3): 350–361.

Game, A. (1991) *Undoing the Social*. Open University Press, Buckingham.

Gaster, L. (1996) 'Decentralisation, Empowerment and Citizenship'. *Local Government Policy Making* 22 (4): 57–64.

Geertz, C. (1983) *Local Knowledge: Further Essays in Interpretive Anthropology*. Basic, New York.

George, V. and Wilding, P. (1985) *Ideology and Social Welfare*. Routledge and Kegan Paul, London.

Gibb, P. (2000) *The Second Clearances*. Land Reform Scotland, Buckie.

Gibbs, G. (1997) 'Seeing Red at Blue-Bloods' Path Ban'. *Guardian*, 27 September: 6.

Giddens, A. (1979) *Central Problems in Social Theory*. Macmillan, Basingstoke.

Giddens, A. (1982) *Profiles and Critiques in Social Theory*. Macmillan, Basingstoke.

Giddens, A. (1984) *The Constitution of Society*. Polity, Cambridge.

Giddens, A. (1985) *The Nation State and Violence*. Polity, Cambridge.

Giddens, A. (1991) *Modernity and Self-Identity*. Polity, Cambridge.

Giddens, A. (1995) *Politics, Sociology and Social Theory*. Polity, Cambridge.

Giddens, A. (1998) *The Third Way*. Polity, Cambridge.

Giddens, A. (2000) *The Third Way and Its Critics*. Polity, Cambridge.

Gilg, A. (1996) *Countryside Planning*, 2nd edition. Routledge, London.

Glasson, J., Therivel, R. and Chadwick, A. (1999) *An Introduction to Environmental Impact Assessment*. UCL Press, London.

Gloucestershire County Council (1995) *Parish Paths Guide*, 2nd edition. GCC, Gloucester.

Goffman, E. (1967) *Interaction Ritual*. Pantheon, New York.

Goffman, E. (1975) *Frame Analyses: An Essay on the Organisation of Experience*. Penguin, Harmondsworth.

Goodenough, A. and Seymour, S. (1999) ' "Active Environmental Citizens" and Constructions of the Countryside'. Paper presented at the 1999 RGS/IBG conference, Leicester, January 1999.

Goodwin, P. (1999) 'The End of Consensus: The Impact of Participatory Initiatives on Conceptions of Conservation and the Countryside in the UK'. *Environment and Planning D* 17 (4): 383–402.

Gorman, A. (2000) 'Otherness and Citizenship: Towards a Politics of the Plural Community', in Bell, D. and Haddour, A. (eds) *City Visions*. Pearson, Harlow.

Gramsci, A. (1971) *Selections from the Prison Notebooks*. Lawrence and Wishart, London.

Granovetter, M. (1985) 'Economic Action and Social Structure: The Problem of Embeddedness'. *American Journal of Sociology* 91 (3): 481–510.

Grant, J. (1994) 'On Some Public Uses of Planning "Theory"'. *Town Planning Review* 65 (1): 59–76.

Grant, M. (1998) 'Human Rights Query', *Planning*, 20 March: 9.

Grant, M. (2000) 'Third Party Right of Appeal Predicted'. *Planning*, 28 January: 1.

Grant, M. (2001) 'Planning System Faces Up to a Human Rights Overhaul'. *Planning*, 5 January: 9–10.

Grant, W. (1990) 'Rural Politics in Britain', in Bodiguel, M. and Lowe, P. (eds) *Rural Studies in Britain and France*. Belhaven, London.

Grant, W. (1995) *Pressure Groups, Politics and Democracy in Britain*. Harvester Wheatsheaf, Hemel Hempstead.

Guardian (1998) 'No Law on Access Says Government'. *Guardian*, 24 July.

Guardian (1999) '350 Years On, Diggers Reclaim Fairways . . .'. *Guardian*, 5 April: 12.

Guardian (2000a) 'French Blockades Spread to Britain'. *Guardian*, 8 September: 3.

Guardian (2000b) 'Britain's Champions of Liberty'. *Guardian*, 2 October: 18.

Guardian (2000c) 'Court Threats to Planning System'. *Guardian*, 29 November: 11.

Guardian (2000d) 'Protest Erupts in Violence'. *Guardian*, 2 May: 1.

Gyford, J. (1991) *Citizens, Consumers and Councils*. Macmillan, Basingstoke.

Habermas, J. (1987) *Knowledge and Human Interests*. Polity, Cambridge.

Habermas, J. (1988) *Legitimation Crisis*. Polity, Cambridge.

Habermas, J. (1994) 'Citizenship and National Identity', in Van Steenbergen, B. (ed.) *The Condition of Citizenship*. Sage, London.

Halfacree, K. (1996) 'Displacing the Rural Idyll', in Watkins, C. (ed.) *Rights of Way: Policy, Culture and Management*. Cassell, London.

Halfacree, K. (1999) 'Anarchy Doesn't Work Unless You Think About It: Intellectual Interpretation and DIY Culture'. *Area* 31 (2): 209–220.

Hall, C. and Jenkins, J. (1995) *Tourism and Public Policy*. Routledge, London.

Hall, P. and Ward, C. (1998) *Sociable Cities*. Wiley, Chichester.

Hall, P., Hebbert, M. and Lusser, H. (1993) 'The Planning Background', in Blowers, A. (ed.) *Planning for a Sustainable Environment*. Earthscan, London.

Hall, S. (1996) 'Who Needs "Identity"?', in Hall, S. and Du Gay, P. (eds) *Questions of Cultural Identity*. Sage, London.

Ham, C. and Hill, M. (1993) *The Policy Process in the Modern Capitalist State*, 2nd edition. Harvester Wheatsheaf, Brighton.

Hanbury-Tenison, R. (1997) 'Life in the Countryside', *Geographical Magazine* 69: 88–95.

Harbrecht, D. (1989) 'The Second Coming of Ralph Nader'. *Business Week*, 16 December: 28.

Harper, K. (1996) 'Ministers Delay Decision on £76m Salisbury Bypass'. *Guardian*, 30 October: 7.

Harrison, C. (1991) *Countryside Recreation*. TMS Partnership, London.

Hart, M. (2000) 'I Feel like a Cheat and a Failure'. *The Ecologist* 30 (4): 33.

Harvey, D. (1989) *The Postmodern Condition*. Blackwell, Oxford.

Hay, C. (1996) *Restating Social and Political Change*. Open University Press, Milton Keynes.

Healey, A., Elson, M. and Donk, A. (1988) *Land Use Planning and the Mediation of Urban Change*. Cambridge University Press, Cambridge.

Healey, P. (1997) *Collaborative Planning*. Macmillan, Basingstoke.

Heater, D. (1990) *Citizenship: The Civic Ideal*. Longman, London.

Heclo, H. (1972) 'Review Article: Policy Analysis'. *British Journal of Political Science* 2 (1): 83–108.

Held, D. (1989) *Political Theory and the State*. Polity, Cambridge.

Hereford Times (1994) 'Shareholders to Back Wye Plan'. *Hereford Times*, 8 December.

Hereford Times (1995) 'Search Launched for Navigation Shares'. *Hereford Times*, 23 February.

Herman, D. (1993) 'The Rights Debate'. *Social and Legal Studies* 2 (1): 25–43.

Hetherington, K. (1996) 'Identity Formation, Space and Social Centrality'. *Theory, Culture and Society* 13 (4): 35–52.

Hewison, R. (1987) *The Heritage Industry*. Methuen, London.

Highways Agency (1996) *A34 Newbury Bypass Newsletter* no. 2 (July). Highways Agency, Birmingham.

Highways Agency (1997) Personal communication marked 'A34 Newbury Bypass Security', regarding costs associated with policing and securing the A34 Newbury Bypass route, 11 November. Highways Agency, Birmingham.

Hill, C. (1996) *Liberty against the Law*. Penguin, Harmondsworth.

Hill, D. (1974) *Democratic Theory and Local Government*. Allen and Unwin, London.

Hill, H. (1980) *Freedom to Roam*. Moorland Press, Ashbourne.

Hilton, R. (1976) *The Transition from Feudalism to Capitalism*. New Left Books, London.

Hobsbawm, E. and Rudé, G. (1973) *Captain Swing*. Penguin, Harmondsworth.

Hodge, I. (1996) 'On Penguins on Icebergs: The Rural White Paper and the Assumptions of Rural Policy'. *Journal of Rural Studies* 12 (4): 331–337.

Hohfeld, W. (1919) *Fundamental Legal Conceptions*. Yale University Press, New Haven, Conn.

Holloway, L. and Kneafsey, M. (2000) 'Reading the Space of Farmers' Markets'. *Sociologia Ruralis* 40 (3): 285–299.

Holt, D. (1998) 'Does Cultural Capital Structure American Consumption?'. *Journal of Consumer Research* 25: 1–25.

Home Office (1997a) Interview with Administration of Justice Group official, 30 September.

Home Office (1997b) Personal communication from the Crime and Criminal Justice Unit, Home Office, dated 13 October.

Home Office (1999) 'Bill to Reform Terrorism Law Published Today.' Press release no. 353/99, accessed 2 December 1999.

Home Office (2000) 'Government to "Bring Rights Home".' Press release accessed on 3 October 2000 on the Home Office website, <www.homeoffice.gov.uk/hract>.

Honore, A. (1961) 'Ownership', in Guest, A. (ed.) *Oxford Essays in Jurisprudence*. Oxford University Press, Oxford.

Hoskins, W. (1963) *Common Lands of England and Wales*. Batsford, London.

Hughes, D. (1996) *Environmental Law*, 3rd edition. Butterworths, Oxford.

Hutton, W. (1996) *The State We're In*. Vintage, London.

Hutton, W. (1997) *The State to Come*. Vintage, London.

Ilbery, B. (ed.) (1998) *The Geography of Rural Change*. Longman, Harlow.

Innes, J. (1996) 'Planning through Consensus-Building: A New View . . .'. *Journal of the American Planning Association* 62 (4): 460–472.

Institute for Citizenship Studies (1999) *Encouraging Citizenship: An Introduction to the Institute for Citizenship Studies*. ICS, London.

Institute for Citizenship Studies (2000) Institute website at <www.citizen.org.uk> (accessed December 2000).

Institute of Citizenship (1992) *Introducing the Institute of Citizenship*. Institute for Citizenship, London.

Isin, E. and Wood, P. (1999) *Citizenship and Identity*. Sage, London.

Jacobs, H. (ed.) (1998) *Who Owns America?* University of Wisconsin Press, Madison.

Jameson, F. (1991) *Postmodernism, or the Cultural Logic of Late Capitalism*. Verso, London.

Jefferies, J. (2000) 'Land for Life and Livelihoods: A Campaign for Land Rights'. *Ecos* 21: 45–48.

Jenkins, R. (1992) *Pierre Bourdieu*. Routledge, London.

Jessop, B. (1990) *State Theory: Putting the Capitalist State in Its Place*. Polity, Cambridge.

Johnston, B. (2000) 'Planners Ready to Tackle Rights Issue'. *Planning*, 29 September: 12–13.

Johnston, R. and Pattie, C. (1998) 'Composition and Context: Region and Voting in Britain Revisited during Labour's 1990s Revival'. *Geoforum* 29 (3): 309–329.

Johnston, R., Pattie, C., Dorling, D., MacAllister, I., Tunstall, H. and Rossiter, D. (2001) 'Social Locations, Spatial Locations and Voting at the 1997 British General Election'. *Political Geography* 20 (1): 74–87.

Jones, D. (1989) 'English Social Protest', in Mingay, G. (ed.) *The Unquiet Countryside*. Routledge, London.

Jones, R. (1999) 'The Mechanics of Medieval State Formation: Observations from Wales'. *Space and Polity* 3 (1): 85–99.

Justice? (1995) *The White Book*, September. Justice?, Brighton.

Justice? (1996) *schNEWS* no. 60, 9 February. Justice?, Brighton.

Kearns, A. (1995) 'Active Citizenship and Local Governance: Political and Geographical Dimensions'. *Political Geography* 14 (2): 155–175.

Keat, R. and Urry, J. (1982) *Social Theory as Science*. Routledge and Kegan Paul, London.

Keeble, J. (1995) 'Hunt, Shoot, Fish, Kill'. *Guardian Society*, 15 November: 5.

Keith, M. and Pile, S. (1993) *Place and the Politics of Identity*. Routledge, London.

Ketola, M. (1997) 'Ecological Eldorado: Eliminating Excess over Ecology', in Welford, R. (ed.) *Hijacking Environmentalism*. Earthscan, London.

Killingray, D. (1994) 'Rights, "Riot" and Ritual: The Knole Park Access Dispute, Sevenoaks, Kent, 1883–5'. *Rural History* 5 (1): 63–79.

King, E. (1987) *The New Right*. Macmillan, Basingstoke.

Kinsman, P. (1996) 'Re-negotiating the Boundaries of Race and Citizenship: The Black Environment Network and Environmental and Conservation Bodies', in Milbourne, P. (ed.) *Revealing Rural Others*. Pinter, London.

Klein, N. (2000) *No Logo*. Flamingo, London.

Klug, F. (1997) *Reinventing Democracy*, accessed online at <www.charter88.org>, 18 September 2000.

Klug, F., Starmer, K. and Weir, S. (1996) 'Civil Liberties and the Parliamentary Watchdog: The Passage of the Criminal Justice and Public Order Act 1994'. *Parliamentary Affairs* 49 (1): 536–549.

Kousis, M. (1998) 'Ecological Marginalisation in Rural Areas: Actors, Impacts, Responses'. *Sociologia Ruralis* 38 (1): 86–108.

Kousis, M. (1999) 'Environmental Protest Cases: The City, the Countryside, and the Grassroots in Southern Europe'. *Mobilisation* 4 (2): 223–238.

Kristeva, J. (1984) *Revolution in Poetic Language*. Columbia University Press, New York.

Kymlicka, W. (1995) *Multicultural Citizenship*. Oxford University Press, Oxford.

Kymlicka, W. and Norman, W. (1994) 'Return of the Citizen: A Survey of Recent Work on Citizenship Theory'. *Ethics* 104 (2): 352–381.

Laclau, E. (1990) *New Reflections on the Revolution of Our Time*. Phronesis, London.

Laclau, E. and Mouffe, C. (1985) *Hegemony and Socialist Strategy: Towards a Radical Democratic Politics*. Verso, London.

Laddie, Justice (1996) Declaration of Mr Justice Laddie in the case of *NRA* v. *Stockinger*, Court of Chancery.

Lash, S. and Urry, J. (1994) *Economies of Signs and Space*. Sage, London.

Laski, H. (1928) *The Recovery of Citizenship*. Ernest Benn, London.

Latacz-Lohmann, U. and Laughton, R. (2000) 'Farmers' Markets in the UK: A Study of Farmers' Perceptions'. *Farm Management* 10 (10): 579–588.

Latour, B. (1987) *Science and Action: How to Follow Scientists and Engineers through Society*. Open University Press, Milton Keynes.

Latour, B. (1994) 'On Technical Mediation: Philosophy, Sociology and Genealogy'. *Common Knowledge* 4: 29–64.

Law, J. (1994) *Organising Modernity*. Blackwell, Oxford.

Law, J. (1999) 'After ANT: Complexity, Naming and Topology', in Law, J. and Hassard, J. (eds) *Actor Network Theory and After*. Blackwell, Oxford.

Law, J. and Hassard, J. (eds) (1999) *Actor Network Theory and After*. Blackwell, Oxford.

Leadbetter, C. and Christie, I. (1999) *To Our Mutual Advantage*. Demos, London.

Lean, G. (1997) 'Blair Faces Rural Revolt over "Right to Roam"'. *The Independent*, 5 October.

Lefebvre, H. (1991) *The Production of Space*. Polity, Cambridge.

Lefebvre, H. (1996) *Writings on Cities*. Blackwell, Oxford.

Lessnoff, M. (1986) *Social Contract*. Macmillan, Basingstoke.

LGMB (Local Government Management Board) (1999) *Guidance on Enhancing Public Participation in Local Government*. Department of the Environment, Transport and the Regions, London.

Liberty (1995) *Defend Diversity, Defend Dissent*. National Council for Civil Liberties, London.

Lichfield, N. (1965) 'Land Nationalisation', in Hall, P. (ed.) *Land Values*. Sweet and Maxwell, London.

Lodge, A. (1999) *An Account of Things*. Personal account of the Diggers March on 1 April 1999, posted on the Diggers350 e-group discussion list, 8 April 1999.

Lowe, P. (1996) 'The Rural White Papers'. *Wildlife and Countryside Link*, December.

Lowe, P. (2001) 'The Rural White Paper: Sifting through the Bran Tub'. *Town and County Planning*, January: 18–19.

Lowe, P. and Goyder, J. (1983) *Environmental Groups in Politics*. Unwin Hyman, London.

Lowe, P., Clark, J. and Cox, G. (1993) 'Reasonable Creatures: Rights and Rationalities in Valuing the Countryside'. *Journal of Environmental Planning and Management* 36 (1): 101–115.

Lowe, R. and Shaw, W. (1993) *Travellers: Voices of the New Age Nomads*. Fourth Estate, London.

Lowenthal, D. (1985) *The Past Is a Foreign Country*. Cambridge University Press, Cambridge.

Lowerson, J. (1980) 'Battles for the Countryside' in Gloversmith, F. (ed.) *Class, Culture and Social Change*. Harvester, Brighton.

Luhmann, N. (1982) *The Differentiation of Society*. Columbia University Press, New York.

McCargo, D. (2000) *Contemporary Japan*. St Martin's Press, New York.

MacHoul, D. and Grace, A. (1995) *A Foucault Primer*. Routledge, London.

McIntosh, M., Leipziger, D., Jones, K. and Coleman, G. (1998) *Corporate Citizenship*. Financial Times/Pitman, London.

Macintyre, C. (1999) 'The Stakeholder Society and the Welfare State: Forward to the Past'. *Contemporary Politics* 5 (2): 121–136.

McKay, G. (ed.) (1998) *DIY Politics*. Verso, London.

McKian, S. (1995) 'That Great Dustheap Called History: Recovering the Spaces of Citizenship'. *Political Geography* 14 (2): 209–216.

McLuhan, M. and Powers, B. R. (1992) *The Global Village: Transformations in World Life and Media in the 21st Century*. Oxford University Press, Oxford.

Macnaghten, P. and Urry, J. (1998) *Contested Natures*. Sage, London.

Macpherson, C. (1962) *The Doctrine of Possessive Individualism*. Cambridge University Press, Cambridge.

MAFF (1972) Correspondence with the Wye River Authority from R. F. Roberts, MAFF, dated 14 April and headed 'Proposed Byelaws – Water Resources Act 1963 s79'.

MAFF (2000) 'A Strategy for Agriculture'. Ministry of Agriculture, Fisheries and Food website, <www.MAFF.gov.uk/farm/agendatwo> (accessed 8 September 2000).

MAFF/DoE (1995) *This Rural England: A Nation Committed to Living Countryside*. Rural White Paper. HMSO, London.

Malatesta, E. (1974) *Anarchy*. Lawrence and Wishart, London.

Malcolmson, J. (1973) *Recreation in the Eighteenth Century*. Cambridge University Press, Cambridge.

Mandelbaum, S. (1991) 'Telling Stories'. *Journal of Planning Education and Research* 10: 209–214.

Mann, M. (1987) 'Ruling Class Strategies and Citizenship'. *Sociology* 21 (3): 339–354.

Marriott, E. (1996) 'Buy Your Own and Stop the Developers'. *The Times*, 11 November.

Marsden, T. (1995) 'Beyond Agriculture? Regulating the New Rural Spaces'. *Journal of Rural Studies* 11 (3): 285–296.

Marsden, T., Munton, R., Flynn, A. and Murdoch, J. (1993) *Constructing the Countryside*. UCL Press, London.

Marsh, D. (ed.) (1998) *Comparing Policy Networks*. Open University Press, Buckingham.

Marshall, T. H. and Bottomore, T. (1992) *Citizenship and Social Class*. Pluto Press, London. (Reprinted edition of the 1950 essay.)

Marston, S. (1995) 'The Private Goes Public: Citizenship and the New Spaces of Civil Society'. *Political Geography* 14 (2): 194–198.

Marston, S. and Staeheli, L. (1994) Guest Editorial: 'Citizenship, Struggle, and Political and Economic Restructuring'. *Environment and Planning A* 26: 840–848.

Martin, S. (1995) 'Partnerships for Local Environmental Action: Observations on the First Two Years of Rural Action for the Environment'. *Journal of Environmental Planning and Management* 38 (2): 149–165.

Marvin, S. and Guy, S. (1997) 'Creating Myths Rather than Sustainability: The Transition Fallacies of the New Localism'. *Local Environment* 2 (3): 311–318.

Massey, D. (1993) 'Politics and Space/Time', in Keith, M. and Pile, S. (eds) *Place and the Politics of Identity*. Routledge, London.

Massey, D. (2000) 'The Geography of Power', accessed through *Red Pepper* website, <www.redpepper.org.uk/xglobal> (July 2000).

Mauss, M. (1990) *The Gift. Forms and Functions of Exchange in Archaic Societies.* Routledge, London.

Merrick, G. (1996) *Battle for the Trees.* Godhaven, Leeds.

MHLG (Ministry of Housing and Local Government) (1942) *Land Utilisation in Rural Areas.* The Scott Report. HMSO, London.

Milbourne, P. (ed.) (1996) *Revealing Rural Others.* Pinter, London.

Miles, S. (1998) *Consumerism.* Sage, London.

Miller, S. (1996) 'Class, Power and Social Constructionism: Issues of Theory and Application in 30 Years of Rural Studies'. *Sociologia Ruralis* 36 (1): 93–116.

Mingay, G. (ed.) (1989) *The Unquiet Countryside.* Routledge, London.

Mingay, G. (1994) *Land and Society in England 1750–1980.* Longman, London.

Minton, A. (1999) 'Swampy Joins Forces with Lawyers and Insurers'. *The Times,* 22 March: 46.

Mintzberg, H. (1983) *Power in and around Organisations.* Prentice-Hall, Englewood Cliffs, NJ.

Mitchell, D. (1992) 'Iconography and Locational Conflict from the Underside'. *Political Geography* 11 (2): 152–169.

Mobbs, P. (2000) 'The Internet: Disintermediation and Campaign Groups'. *Ecos* 21 (1): 25–32.

Mol, A. and Law, J. (1994) 'Regions, Networks and Fluids: Anaemia and Social Topology'. *Social Studies of Science* 24: 641–671.

Molotch, H. (1999) 'All Industries Are Culture Industries'. Paper given at the University of Surrey, Department of Sociology seminar series, April 1999.

Monbiot, G. (1998) 'Reclaim the Fields and Country Lanes: The Land Is Ours Campaign', in McKay, G. (ed.) *DIY Culture*. Verso, London.

Monbiot, G. (1999) 'Conservation by Rights', in Barrett, A. and Scruton, R. (eds) *Town and Country*. Vintage, London.

Monbiot, G. (2000) 'Welcome to Britain plc'. *Guardian Saturday Review*, 9 September: 1.

Money Programme (1996) 'Protests and Profits'. *The Money Programme*, TV programme shown on BBC 2, 24 March.

Mormont, M. (1987) 'The Emergence of Rural Struggles and Their Ideological Effects'. *International Journal of Urban and Regional Research* 7: 559–578.

Mormont, M. (1990) 'Who Is Rural? Or, How to Be Rural', in Marsden, T., Lowe, P. and Whatmore, S. (eds) *Rural Restructuring*. David Fulton, London.

Morton, J. (1994) *A Guide to the Criminal Justice and Public Order Act 1994.* Butterworths, London.

Moseley, M. (2000) 'Innovation and Rural Development: Some Lessons from Britain and Western Europe'. *Planning Practice and Research* 15: 95–115.

Moseley, M. and Parker, G. (1998) *The Joint Provision of Rural Services.* Rural Research Report no. 34. Rural Development Commission, Salisbury.

Moseley, M., Derounian, J. and Allies, P. (1996) 'Parish appraisals – A Spur to Local Action? A Review of the Gloucestershire and Oxfordshire Experience 1990–1994'. *Town Planning Review* 67 (3): 309–330.

Mouffe, C. (1993) 'Liberal Socialism and Pluralism: Which Citizenship?', in Squires, J. (ed.) *Principled Positions*. Lawrence and Wishart, London.

Mouffe, C. (1995) 'Post-Marxism: Democracy and Identity'. *Environment and Planning D: Society and Space* 13: 259–265.

Munton, R. (1994) 'Rural Accumulation and Property Rights: Sustaining the Means'. Paper presented at the IBG annual conference, Nottingham, January 1994.

Munton, R. (1995) 'Regulating Rural Change: Property Rights, Economy and Environment. A Case Study from Cumbria, UK'. *Journal of Rural Studies* 11 (3): 269–284.

Murdoch, J. (1997a) 'The Shifting Territory of Government: Some Insights from the Rural White Paper'. *Area* 29 (2): 109–118.

Murdoch, J. (1997b) 'Tracing the Topologies of Power'. Unpublished paper presented at the ANT and After Conference, University of Keele, July 1997. Located at <http//www.keele.ac.uk/depts/stt/cstt.htm>.

Murdoch, J. (1998) 'The Spaces of Actor–Network Theory'. *Geoforum* 29 (4): 357–374.

Murdoch, J. (1999) 'The New Rural White Paper: What Chance for a Rural Policy?'. *Countryside Recreation* 7 (1): 10–11.

Murdoch, J. and Abram, S. (1998) 'Defining the Limits of Community Governance'. *Journal of Rural Studies* 14 (1): 41–50.

Murdoch, J. and Marsden, T. (1994) *Reconstituting Rurality*. UCL Press, London.

Murdoch, J. and Pratt, A. (1993) 'Rural Studies: Modernism, Postmodernism and the 'Post-rural'. *Journal of Rural Studies* 9 (4): 411–427.

Murdoch, J. and Pratt, A. (1994) 'Rural Studies and Power and the Power of Rural Studies: A Reply to Philo'. *Journal of Rural Studies* 10 (1): 83–87.

Murdoch, J. and Pratt, A. (1997) 'From the Power of Topography to the Topography of Power: A Discourse on Strange Ruralities', in Cloke, P. and Little, J. (eds) *Contested Countryside Cultures*. Routledge, London.

Nader, R. (1990) *Auto Rights*. Moyer Bell, New York.

National Policy Forum (1999) *National Policy Forum Draft Document*. June 1999. Labour Party, London.

Negrine, R. (1996) *The Communication of Politics*. Sage, London.

Nevin, C. (1997) 'Of Sprouts and Men'. *Guardian*, 17 April: 5.

New Economics Foundation (1999) *Community Works!* NEF, London.

Newby, H. (1979) *Green and Pleasant Land*. Hutchinson, London.

Newby, H. (1987) *Country Life*. Penguin, London.

Newby, H. (1996) 'Citizenship in a Green World', in Bulmer, M. and Rees, A. (eds) *Citizenship Today*. UCL Press, London.

Newby, H., Bell, H., Rose, D. and Saunders, P. (1978) *Property, Paternalism and Power*. Hutchinson, London.

Norton, A. (2000) 'The Relationship between Hunting and Country Life'. *Countryside Recreation* 8 (1): 8–11.

Norton-Taylor, R. (1982) *Whose Land Is It Anyway?* Turnstone Press, London.

Nowotny, H. (1994) *Time*. Polity, Cambridge.

NRA (National Rivers Authority) (1995) *Catchment Management Plan for the Wye Valley*. NRA, Cardiff.

Observer (1998) 'The Rural Goes to War'. *Observer*, 22 February: 6.

Observer (2000) 'Fuel Crisis Hits Home'. *Observer*, 17 September.

Oliver, D. (1991) 'Active Citizenship in the 1990s'. *Parliamentary Affairs* 44 (2): 157–171.

Opinion Leader Research (1996) Unpublished Brent Spar research report conducted on behalf of Greenpeace, January. Opinion Leader Research, London.

Out of This World (1997) *World News*, no. 6 (October): 20.

PACEC (PA Cambridge Economic Consultants) (1995) *Final Evaluation Report on the Parish Paths Partnership Scheme*. Unpublished report, November 1995. PACEC, Cambridge.

Pahl, R. (1998) 'Friendship: The Social Glue of Contemporary Society?' in Franklin, J. (ed.) *The Politics of Risk Society*. Polity, Cambridge.

Painter, J. and Philo, C. (1995) 'Spaces of Citizenship: An Introduction'. *Political Geography* 14 (2): 107–120.

Parker, G. (1995) 'Access Liaison Groups: An Assessment'. Unpublished report to the Countryside Commission, Cheltenham, October.

Parker, G. (1996) 'ELMs Disease: Stewardship, Corporatism and Citizenship in the English Countryside'. *Journal of Rural Studies* 12 (4): 399–411.

Parker, G. (1997) 'Citizens' Rights and Private Property Rights in the English Countryside'. Unpublished Ph.D. thesis, University of Bristol.

Parker, G. (1999a) 'Rights, the Environment and Part V of the Criminal Justice and Public Order Act 1994'. *Area* 31: 75–80.

Parker, G. (1999b) 'Rights, Symbolic Violence and the Micro-politics of the Rural: The Case of the Parish Paths Partnership Scheme'. *Environment and Planning A* 31: 1207–1222.

Parker, G. (1999c) 'The Role of the Consumer-Citizen in Environmental Protest in the 1990s'. *Space and Polity* 3 (1): 67–83.

Parker, G. (2000) 'Governing Access? Countryside Access Liaison Groups: A Caution . . .'. *Countryside Recreation* 8 (2): 12–17.

Parker, G. (2001) 'Rights and Planning: Some Repercussions of the Human Rights Act 1998'. *Planning Practice and Research* 16 (1): 5–8.

Parker, G. and Ravenscroft, N. (1999) 'Benevolence, Nationalism and Hegemony: Fifty Years of the National Parks and Access to the Countryside Act 1949'. *Leisure Studies* 18: 297–313.

Parker, G. and Ravenscroft, N. (2001) 'CROW 2000, Citizenship and the Gift'. Paper given at the RESSG seminar Reconceptualising Property Rights, University of Reading, March 2001.

Parker, G. and Wragg, A. (1999) 'Networks, Agency and (De)stabilization: The Issue of Navigation on the River Wye, UK'. *Journal of Environmental Planning and Management* 42 (4): 471–488.

Passerin-d'Entreves, M. (1994) *The Political Philosophy of Hannah Arendt*. Routledge, London.

Pedder, C. (1999) '"Diggers" Beat the Bailiffs'. *Surrey Comet*, 16 April.

Peet, R. (1998) *Modern Geographical Thought*. Blackwell, Oxford.

Penning-Rowsell, E. (1994) 'A "Tragedy of the Commons"? Perceptions of Managing Recreation on the River Wye, UK'. *Natural Resources Journal* 34 (Summer): 629–655.

Penning-Rowsell, E. (1996) *Proof of Evidence to the Wye Navigation Order Public Local Inquiry*. Environment Agency, Cardiff.

Penning-Rowsell, E. and Crease, D. (1988) 'Water for Amenity and Recreation: Legal Constraints on Planning and Management for the River Wye'. *Landscape and Urban Planning* 16: 105–125.

Petegorsky, D. (1995) *Left-Wing Democracy in the English Civil War*. Gollancz, London.

Philips, D. (1993) *Looking Backwards: A Critical Appraisal of Communitarian Thought*. Princeton University Press, Princeton, NJ.

Pickerill, J. (2000) 'Spreading the Green Word? Using the Internet for Environmental Campaigning'. *Ecos* 21 (1): 14–24.

Pilcher, J. and Wagg, S. (1996) *Thatcher's Children?* Falmer Press, London.

PIU (Performance and Innovation Unit) (1999) *Rural Economies*. Cabinet Office, London.

Plant, V. (1994) 'Democratic Citizenship', in Clarke, P. (ed.) *Citizenship: A Reader*. Pluto Press, London.

Potter, J. (1988) 'Consumerism and Public Sector: How Well Does the Coat Fit?'. *Public Administration* 66 (2): 149–164.

Poulantzas, N. (1968) *Political Power and Social Classes*. New Left Books, London.

Pred, A. (1984) 'Place as Historically Contingent Process'. *Annals of the Association of American Geographers* 74 (2): 279–297.

Pritchard, R. (2000) 'The Human Rights Act 1998: Implications for Planning'. *Town and Country Planning*, October: 284–285.

Private Investigations (1997) Feature on Bray's Investigation Agency. TV programme shown on BBC TV, 9 October.

Prott, A. (1998) *Human Rights*. UNESCO, Geneva.

Public Accounts Committee (1999) 'BSE: The Cost of a Crisis'. PAC Report no. 34, July. Public Accounts Committee, HMSO, London.

Raban, J. (1988) *Soft City*. Collins, London.

Rabinow, P. (1991) *The Foucault Reader*. Penguin, Harmondsworth.

Radcliffe, S. (1993) 'Women's Place. Latin America and the Politics of Gender Identity', in Keith, M. and Pile, S. (eds) *Place and the Politics of Identity*, Routledge, London.

Ramblers' Association (1991) *Who Speaks for Us?* Ramblers' Association, London.

Ramblers' Association (1998) '"Keep Your Promise, Tony Blair", Ramblers Tell Prime Minister'. Press release dated 25 February, Ramblers' Association, London.

Ravenscroft, N. (1993) 'Public Leisure Provision and the Good Citizen'. *Leisure Studies* 12 (1): 33–44.

Ravenscroft, N. (1998) 'Rights, Citizenship and Access to the Countryside'. *Space and Polity* 2 (1): 33–48.

Ravenscroft, N. (1999) 'Hyperreality and Rambling', in Crouch, D. (ed.) *Leisure/Tourism Geographies*. Routledge, London.

Ravenscroft, N. and Parker, G. (1999) 'Regulating Time, Regulating Society'. *Loisir et Société* 22: 381–401.

Ravenscroft, N., Markham, S. and Curry, N. (1996) 'Evaluation of Highways Authority Milestones Statements'. Unpublished report to the Countryside Commission, Cheltenham. CELTS, University of Reading.

Rawls, J. (1979) *A Theory of Justice*. Oxford University Press, Oxford.

Ray, C. (1999) 'Towards a Meta-framework of Endogenous Development: Repertoires, Paths, Democracy and Rights'. *Sociologia Ruralis* 39 (4): 521–537.

Rhodes, R. (1988) *Beyond Westminster and Whitehall*. Routledge, London.

Rhodes, R. (1997) *Rethinking Governance*. Routledge, London.

Riddall, J. and Trevelyan, J. (1992) *Rights of Way: A Guide to Law and Practice*, 2nd edition. Open Spaces Society/Ramblers' Association, London.

Ritzer, G. (1996) *Sociological Theory*. McGraw-Hill, New York.

River Wye Project (1992) *Conservation and Recreation: The Wye Challenge*. The River Wye Project, Monmouth.

Robertson, R. (1995) 'Glocalization: Time–Space and Homogeneity–Heterogeneity', in Featherstone, M. (ed.) *Global Modernities*. Sage, London.

Roche, M. (1992) *Rethinking Citizenship*. Polity, Cambridge.

Rogers, A. (1987) 'Voluntarism, Self-Help and Rural Community Development: Some Current Approaches'. *Journal of Rural Studies* 3 (4): 353–360.

Rojek, C. and Urry, J. (1997) *Touring Cultures. Transformations in Travel and Theory*. Routledge, London.

Rose, C. (1996) 'Whose Responsibility Is It Anyway?'. Speech made at the Corporate Brand Conference, London, 15 February 1996.

Rose, C. (1997) 'Why "Solutions"?'. Paper delivered at the Marine Environmental Management Conference, University of London, 23 January 1997.

Rose, N. (1990) *Governing the Soul: The Shaping of the Private Self*. Routledge, London.

Rose, N. (1996) 'Governing "Advanced" Liberal Democracies', in Barry, A., Osborne, T. and Rose, N. (eds) *Foucault and Political Reason*. UCL Press, London.

Ross Gazette (1994) 'Franks' Dream for the Wye'. *Ross Gazette*, 8 September: 6.

Rothman, B. (1982) *The 1932 Kinder Trespass*. Willow Press, Altrincham.

Routledge, P. (1996) 'The Third Space as Critical Engagement'. *Antipode* 28 (4): 399–419.

Routledge, P. (1997) 'The Imagineering of Resistance: Pollok Free State and the Practice of Postmodern Politics'. *Transactions of the Institute of British Geographers*, NS 22: 359–376.

Sabatier, P. (1987) 'Knowledge, Policy Oriented Learning and Policy Change: An Advocacy Coalition Framework'. *Knowledge: Creation, Diffusion, Utilization* 1: 649–692.

Sabine, G. (ed.) (1965) *The Collected Works of Gerrard Winstanley*. Russell and Russell, London.

Sabine, G. and Thorson, T. (1973) *A History of Political Theory*, 4th edition. Holt Saunders, Tokyo.

Sack, R. (1986) *Human Territoriality: Its Theory and History*. Cambridge University Press, Cambridge.

Sack, R. (1993) *Place, Modernity and the Consumers World*. Johns Hopkins University Press, Baltimore.

Sagoff, M. (1988) *The Economy of the Earth*. Cambridge University Press.

Saul, J. (1992) *Voltaire's Bastards*. Free Press, New York.

Savage, M., Barlow, J., Dickens, P. and Fielding, T. (1992) *Property, Bureaucracy and Culture*. Routledge, London.

Scottish Executive (1998) *Land Reform Policy Group: Identifying the Problems*. Scottish Office, Edinburgh.

Scottish Executive (1999) *Land Reform Policy Group: Recommendations for Action*. The Scottish Office, Edinburgh.

Sedley, S. (1997) 'Human Rights: A 21st Century Agenda', in Blackburn, R. and Busuttil, J. (eds) *Human Rights for the 21st Century*. Pinter, London.

Seel, B., Paterson, M. and Doherty, B. (eds). (2000) *Direct Action in British Environmentalism*. Routledge, London.

Selman, P. (1996) *Local Sustainability*. Paul Chapman Press, London.

Selman, P. (2000) *Environmental Planning*, 2nd edition. Sage, London.

Selman, P. and Wragg, A. (1997) 'Applying the Theory of the Sociology of Translation to the Study of Consensus Building in the UK'. Draft working paper, Countryside and Community Research Unit.

Selman, P. and Wragg, A. (1999) 'Networks of Co-operation and Knowledge in "Wider Countryside" Planning'. *Journal of Environmental Planning and Management* 42 (5): 649–669.

Serres, M. and Latour, M. (1995) *Conversations on Science, Culture and Time.* University of Michigan Press, Ann Arbor, Ill.

Sharp, T. (1932) *Town and Countryside.* Oxford University Press, Oxford.

Shields, R. (1991) *Places on the Margin.* Routledge, London.

Shoard, M. (1980) *The Theft of the Countryside.* Temple Smith, London.

Shoard, M. (1987) *The Land Is Ours.* Grafton, London.

Shoard, M. (1996) 'Robbers v. Revolutionaries: What the Battle for Access Is Really All About', in Watkins, C. (ed.) *Rights of Way: Policy, Culture and Management.* Pinter, London.

Shoard, M. (1999) *A Right to Roam.* Grafton, London.

Shoard, M. (2000) Paper given at Rural History Conference, Reading University, September 2000.

Short, J. (1991) *Imagined Country.* Routledge, London.

Shotter, J. (1993) *Cultural Politics of Everyday Life.* Open University Press, Buckingham.

Sibley, D. (1992) 'Outsiders in Space and Society', in Gale, F. and Anderson, K. (eds) *Inventing Places.* Longman Cheshire, Melbourne.

Sibley, D. (1995) *Geographies of Exclusion.* Routledge, London.

Sibley, D. (1997) 'Endangering the Sacred: Nomads, Youth Cultures and the English Countryside', in Cloke, P. and Little, J. (eds) *Contested Countryside Cultures.*

Simmie, J. (1974) *Citizens in Conflict.* Hutchinson, London.

Simons, J. (1995) *Foucault and the Political.* Routledge, London.

Simpson, A. (1986) *A History of Land Law.* Oxford University Press, Oxford.

Skeffington Report (1969) *Report of the Committee on Public Participation in Planning.* Ministry of Housing and Local Government. HMSO, London.

Sleep, K. (1997) 'Costain at One with Nature on Newbury Bypass'. *Construction News,* 20 March.

Smith, R. (2001) 'Citizenship and the Politics of People-Building'. *Citizenship Studies* 5 (1): 73–96.

Smith, S. (1989) 'Society, Space and Citizenship: A Human Geography for the "new times"?'. *Transactions of the Institute of British Geographers,* NS 14 (2): 144–156.

Smith, S. (1995) 'Citizenship: All or Nothing?'. *Political Geography* 14 (2): 190–193.

Soja, E. (1996) *Thirdspace.* Blackwell, Oxford.

Star, S.-L. (1995) 'The Politics of Formal Representations: Wizards, Gurus and Organisational Complexity', in Star, S.-L. (ed.) *Ecologies of Knowledge.* State University of New York, NY.

Stephenson, T. (1989) *Forbidden Land.* Manchester University Press, Manchester.

Stevenson, N. (ed.) (2001) *Culture and Citizenship.* Sage, London.

Stewart, J. (1995) 'A Future for Local Authorities as Community Government', in Stewart, J. and Stoker, G. (eds) *Local Government in the 1990s.* Macmillan, Basingstoke.

Stewart, J. and Stoker, G. (eds) (1995) *Local Government in the 1990s.* Macmillan, Basingstoke.

Stockinger, V. (1997) *The Rivers Wye and Lugg Navigation: A Documentary History*. Logaston Press, Plymouth.

Stoker, G. (1988) *The Politics of Local Government*. Macmillan, Basingstoke.

Stokols, D. and Shumaker, S. A. (1981) 'People in Places: A Transactional View of Settings', in Harvey, J. (ed.) *Cognition, Social Behavior and the Environment*. Lawrence Erlbaum, Hillsdale, NJ.

Strathern, M. (1999) 'What Is Intellectual Property After?', in Law, J. and Hassard, J. (eds) *Actor Network Theory and After*. Blackwell, Oxford.

Surrey Comet (1999) 'New Levellers Set Up at St George's'. *Surrey Comet*, 5 April: 7.

Surrey Comet (2000) 'St George's Diggers Are Honoured at Last'. *Surrey Comet*, 24 March.

Tait, M. and Campbell, H. (2000) 'The Politics of Communication between Planning Officers and Politicians'. *Environment and Planning A* 32 (3): 489–506.

Tannahill, R. (1975) *Food in History*. Paladin Press, St Albans.

Tawney, R. (1926) *Religion and the Rise of Capitalism*. Penguin, Harmondsworth.

Taylor, C. (1995) *Philosophical Arguments*. Harvard University Press, Cambridge, Mass.

Taylor, P. (1993) *Political Geography*, 3rd edition. Longman, Harlow.

Thirsk, J. (1967) *The Agrarian History of England and Wales*. Cambridge University Press, Cambridge.

Thirsk, J. (1999) *Alternative Agriculture*. Oxford University Press, Oxford.

Thompson, E. (1973) *Whigs and Hunters: The Origins of the Black Acts*. Penguin, London.

Thompson, E. (1975) *Origins of the English Working Class*. Penguin, Harmondsworth.

Thompson, E. (1993) *Customs in Common*. Penguin, London.

Thornley, A. (1993) *Urban Planning after Thatcherism*, 2nd edition. Routledge, London.

Thrift, N. (1983) 'On the Determination of Social Action in Space and Time'. *Environment and Planning D: Society and Space* 1: 23–57.

Thrift, N. (1996) *Spatial Formations*. Sage, London.

Thrift, N. (1999) 'The Place of Complexity'. *Theory, Culture and Society* 16 (3): 31–69.

Throgmorton, J. (1992) 'Planning as Persuasive Storytelling about the Future: Negotiating an Electric Power Rate Settlement in Illinois'. *Journal of Planning Education and Research* 12: 17–31.

Tindale, S. (1998) 'Procrastination, Precaution and the Global Gamble', in Franklin, J. (ed.) *Politics of Risk Society*. Polity, Cambridge.

TLIO (1994) *The Land Is Ours: A Statement of Aims*. The Land Is Ours, Oxford.

TLIO (1995) *The Land Is Ours*. *Newsletter* no. 1, 24 August. The Land Is Ours, Oxford.

TLIO (1997) *The Land Is Ours*. *Newsletter* no. 11, Winter 1997/98. The Land Is Ours, Oxford.

TLIO (1998) 'Diggers350: – shaping the celebrations'. *The Land Is Ours*. *Newsletter* no. 13, Summer. The Land Is Ours, Oxford.

TLIO (1999) 'Diggers350'. *The Land Is Ours*. *Newsletter* no. 16, Summer. The Land Is Ours, Oxford.

Tomorrow's Company (2000) Website at <www.tomorrowscompany.org.uk> (accessed 10 September 2000).

Tönnies, F. (1963) *Gemeinschaft und Gesellschaft* [Community and Association], trans. C. P. Loomis. Routledge and Kegan Paul, London.

Trevelyan, J. (1967) *English Social History.* Penguin, Harmondsworth.

Tuan, Y.-F. (1974) *Topophilia: A Study of Environmental Perception, Attitudes and Values.* Prentice-Hall, Englewood Cliffs, NJ.

Turnbull, A. (1996) 'Road Protesters Take to the Trees'. *Salisbury Journal,* 17 October: 1.

Turner, B. (1986) *Citizenship and Capitalism.* Allen and Unwin, London.

Turner, B. (1994) 'Postmodern Culture/Modern Citizenship', in Van Steenbergen, B. (ed.) *The Condition of Citizenship.* Sage, London.

Turner, M. (1980) *English Parliamentary Enclosures.* Archon, Folkestone.

Urry, J. (1995) *Consuming Places.* Routledge, London.

Urry, J. (2000) *Sociology beyond Societies: Mobilities for the Twenty-first Century.* Routledge, London.

Uzzell, D., Groeger, J., Ravenscroft, N. and Parker, G. (2000) 'User Interaction on Multi-use Routes'. Unpublished report to the Countryside Agency, July 2000. (Précis available on the Countryside Agency website, <www.countryside.gov.uk>.)

Van Gunsteren, H. (1994) 'Four Conceptions of Citizenship', in Van Steenbergen, B. (ed.) *The Condition of Citizenship.* Sage, London.

Van Gunsteren, H. (1998) *A Theory of Citizenship.* Westview Press, Oxford.

Van Steenbergen, B. (ed.) (1994) *The Condition of Citizenship.* Sage, London.

Vidal, J. (2000) 'Ben Threw the First Stone and . . .'. *The Guardian,* 2 May: 3.

Vincent, A. and Plant, R. (1984) *Philosophy, Politics and Citizenship.* Blackwell, Oxford.

Waddington, P. (1994) *Liberty and Order.* UCL Press, London.

Ward, C. (1978) *The Child in the Country.* Architectural Association, London.

Ward, C. (1999) 'The Unofficial Countryside', in Barnett, A. and Scruton, R. (eds) *Town and Country.* Vintage, London.

Warde, A. (1994) 'Consumers, Consumption and Post-Fordism', in Burrows, R. and Loader, B. (eds) *Towards a Post-Fordist Welfare State?* Routledge, London.

Weale, S. (1995) 'Travellers' Win Makes Law Unworkable'. *Guardian,* 1 September: 5.

Weare, A. (1997) Proof of evidence submitted to the *NRA* v. *Stockinger* case, February. NRA, Cardiff.

Welford, R. (ed.) (1997) *Hijacking Environmentalism.* Earthscan, London.

Welsh, E. (1997) 'Gun Lobby Funds Steel's Countryside Movement'. *Observer,* 12 January: 21.

Western Mail (1995) 'Battle over £85m River Plan'. *Western Mail,* 5 July.

Whatmore, S., Munton, R. and Marsden, T. (1990) 'The Rural Restructuring Process: Emerging Divisions of Agricultural Property Rights'. *Regional Studies* 24 (3): 235–245.

Whitt, A. (1979) 'Towards a Class Dialectical Model of Power'. *American Sociological Review* 44: 81–100.

Wilcox, D. (1994) *The Guide to Effective Participation.* Partnership Books, Brighton.

Wilkinson, J. (1997) 'A New Economic Paradigm for Economic Analysis?'. *Economy and Society* 26 (3): 305–339.

Williams, R. (1973) *The Country and the City.* Chatto and Windus, London.

Winter, M. (1996) *Rural Politics.* Routledge, London.

Woods, M. (1997a) 'Researching Rural Conflicts: Hunting, Local Politics and Actor-Networks'. *Journal of Rural Studies* 14 (3): 321–340.

Woods, M. (1997b) 'From Rural Politics to a Politics of the Rural . . .'. Paper given at the Annual RESSG Conference, Worcester, September 1997.

Woods, M. (1998) 'The People of England Speak? . . .'. Paper given at the Annual RESSG Conference, University College of Wales, Aberystwyth, September 1998.

Woodspring District Council (1994) *Draft Countryside Strategy 1994.* Woodspring Borough Council, Weston-super-Mare.

Wright, A. (1994) *Citizens and Subjects.* Routledge, London.

Wright, P. (1983) *On Living in an Old Country.* Random Century, London.

Wright, P. (1991) *A Journey through Ruins.* Radius, London.

Wright, P. (1996) *The Village that Died for England.* Vintage, London.

Wye Forum (1994) Unpublished minutes of Wye Forum Meetings, 1994. Wye Valley AONB, Monmouth.

Wye Mag (1995) Minutes of the Wye Mag Meeting 31 July. AONB Office, Monmouth.

Wye Preservation Society (1998) *Proof of Evidence to the 1996 Wye Navigation Order Inquiry.* WPS, Hereford.

Wye Valley AONB (1990) *A Strategy for Sustainable Tourism.* Wye Valley Area of Outstanding Natural Beauty Joint Advisory Committee, Monmouth, Gwent.

Wye Valley AONB (1992) *Wye Valley Management Plan.* Gwent County Council, Monmouth.

Yarwood, R. and Gardner, G. (2000) 'Fear of Crime, Cultural Threat and the Countryside'. *Area* 32 (4): 403–411.

Index